MEDIA MANUALS

TV Light

Second Ec

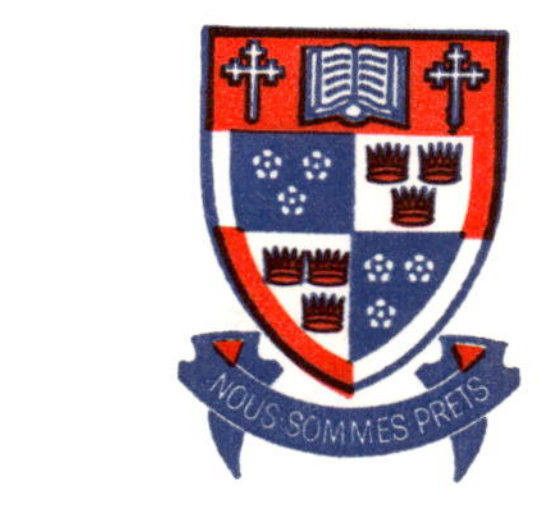

MEDIA MANUALS

VIDEO CAMERA TECHNIQUES
Gerald Millerson

USING VIDEOTAPE
J. J. Robinson & P. H. Beards

BASIC TV STAGING
Gerald Millerson

THE USE OF MICROPHONES
Alec Nisbett

TV SOUND OPERATIONS
Glyn Alkin

TV LIGHTING METHODS
Gerald Millerson

EFFECTIVE TV PRODUCTION
Gerald Millerson

THE SMALL TV STUDIO
Alan Bermingham, Michael Talbot-Smith, John Symons & Ken Angold-Stephens

16-mm FILM CUTTING
John Burder

THE LENS & ALL ITS JOBS
Sidney F. Ray

BASIC FILM TECHNIQUE
Ken Daley

YOUR FILM & THE LAB
L. Bernard Happé

MOTION PICTURE FILM PROCESSING
Dominic Case

MOTION PICTURE CAMERA DATA
David W. Samuelson

MOTION PICTURE CAMERA & LIGHTING EQUIPMENT
David W. Samuelson

MOTION PICTURE CAMERA TECHNIQUES
David W. Samuelson

SCRIPT CONTINUITY & THE PRODUCTION SECRETARY
Avril Rowlands

THE ANIMATION STAND
Zoran Perisic

CREATING SPECIAL EFFECTS FOR TV & FILMS
Bernard Wilkie

LOCAL RADIO
Barrie Redfern

TV Lighting Methods

Second Edition

Gerald Millerson

FOCAL PRESS
London – Boston

Focal Press
is an imprint of the Butterworth Group
which has principal offices in
London, Boston, Durban, Singapore, Sydney, Toronto, Wellington

First published 1975
Second edition 1982
Reprinted 1983, 1986

British Library Cataloguing in Publication Data

Millerson, G.
TV lighting methods – 2nd ed.
1. Television broadcasting – Lighting
I. Title
791.45'025 PN1992.8.L5

ISBN 0-240-51181-6

Photoset by Butterworths Litho Preparation Department
Printed in England by The Whitefriars Press Ltd., Tonbridge

Contents

Introduction

Why on earth should we bother with the complications and expense of special lighting systems, when the camera works with the light already available? If light is needed, why not set up sufficiently powerful lamps near the camera, and flood the scene with light? Where's the problem?

The answers to these questions are what this book is all about! To achieve really effective, persuasive pictures, the TV subject and its surroundings need to be lit systematically, to suit the TV camera. Leave it all to chance, and the production's impact, appeal, and significance also become a matter of luck. And, for most programme purposes, that simply is not good enough.

The craft of lighting is an adroit blend of artistic sensitivity and practical know-how. This requires study and experience to acquire, but that takes time. What if we are concerned only with light as one contributory aspect of the job in hand, and haven't the time or inclination to make a study in depth? That is where this book can help you. It has been designed to enable you to light for your particular application.

You will find included here not only the principles underlying the techniques but also details of the most useful practical approaches. Scaled lighting plots show the commonest set-ups encountered, and typical professional solutions to everyday problems.

Excellent lighting equipment of many types is available throughout the world. Exactly which designs prove most suitable for your own purpose, must depend on availability, and your particular applications. For example, lightweight gear has the advantages of being extremely portable and compact. But if you anticipate regular studio usage, frequent transportation, then you need heavier, rugged equipment that withstands continual rough handling, yet still remains safe and reliable.

You will enjoy lighting. It is a satisfying, creative experience. If you want to know more of the subject, further study sources are listed on page 138 of this book.

Acknowledgement

The author would like to thank the Director of Engineering of the British Broadcasting Corporation for permission to publish this book.

The TV Camera Interprets

It is very easy to assume that our TV camera is just an 'extension electronic eye'. But such an over-simplification is extremely misleading. As our eyes look around freely with stereoscopic colour vision, we form a co-ordinated impression of our surroundings. The camera cannot do this. It can present only a segmented *flat* image of the scene before it – and this picture is often monochromatic.

Translating colour

In a black-and-white picture, all our information about the scene has to be interpreted from varying shades of grey. Perspective clues help us to build up ideas of distance and size, but these can easily become confused in the two-dimensional image. Differing hues can reproduce as identical grey values. Nevertheless, we must ensure that the scene is reproduced clearly and effectively. Often this can be done only by careful systematic lighting that gives appropriate emphasis to form and texture, revealing space and distance. Casual or inappropriate illumination cannot do these things in the flat picture. Even in a colour system we have still to consider those who view in monochrome.

The colour TV camera enables us to distinguish between various hues – although perhaps not accurately. But here again, the quality and the pictorial effect can be influenced considerably by our lighting treatment (page 38).

The camera tube's limitations

The TV camera *pick-up tube*, can handle only a relatively restricted range of tones. Any that are brighter or darker than these limits will not reproduce accurately. Instead, we shall see blank 'burned-out' highlights, and solid detailless black shadows. Fortunately, good staging, compensatory lighting techniques (page 12) and skilled video control (page 132) help to minimise the effects of these restrictions.

The TV camera has other technical limitations, too, such as its responses to excess or insufficient illumination. The series of distracting picture defects that these can produce may be reduced by appropriate staging and lighting (page 24).

The brain makes allowances

Finally, whereas our eyes and brain continually interpret the scene and adjust to variations in its brightness, the camera cannot do this. Our eyes discern detail in shadows, and accommodate to other surroundings, but the camera simply registers what is before it, within its limits. We have to control its *exposure* to suit prevailing conditions, and to ensure that the features we are particularly interested in are clearly and suitably shown. So we must control staging and lighting to suit the camera, or accept the consequences of not doing so.

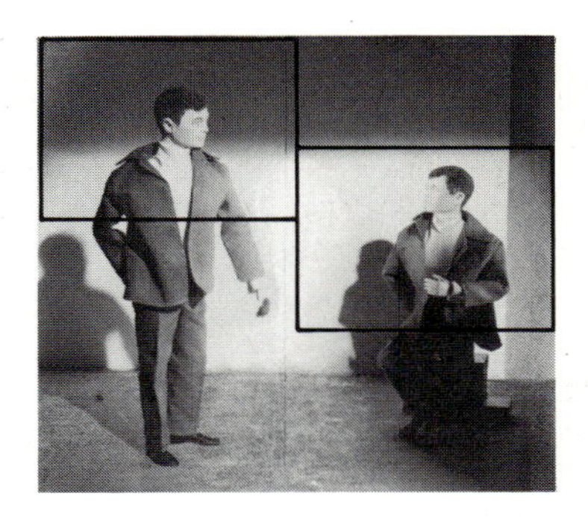

THE TV PICTURE

Field of view
The camera sees a rectangular portion of the scene, in 4-by-3 proportions. The viewers' impressions derive from what they see in this restricted area.

The camera is selective
The wall shading that is effective in the standing shots, is unseen when the person sits. To the camera, it is no longer there.

Why Use Special Lighting?

During daylight hours we have natural light. After dark we normally have interior illumination available. Surely, the light is already there!

Available light
That's true. We can use available light. The results may be excellent . . . provided that the light happens to be right for our purpose, or we shoot to suit the prevailing light, or we are satisfied with the pictures despite their shortcomings.

Daylight, as we shall see (page 112) is extremely variable. Interior illumination takes so many forms, from fluorescent-bank ceilings to localised pools of light. So, even assuming that this lighting is appropriately located for normal environmental activities (and it may not be!), this is no assurance that the illumination will be suitable for our TV cameras, and the shots they are taking. Ugly effects that we disregard on the spot can appear grotesque on the screen.

Lighting to control technical quality
Technically, appropriate lighting ensures that the general level of illumination, and the tonal range of the scene, fall within the camera-tube's limits (pages 10, 132). Light levels affect picture quality and clarity (page 76).

Strategic lighting can influence tonal contrast, increasing or reducing it to suit the occasion. Diffused overall lighting tends to produce minimum contrast. But by systematically illuminating darker tones more strongly, while keeping light off the lightest areas, we may be able, in certain circumstances, to reduce this contrast further. Where these strategies are impracticable, tonal extremes will *crush-out* to plain unmodelled areas.

Artistic control
Artistically, specific lighting enables us to control the *appearance* of subject and scene. Under various lighting treatments, the picture can become attractive or unattractive, effective or valueless for its purpose. The same light conditions cannot suit all situations. Lighting that is decorative or mysterious, for example, may be ineffectual when demonstrating machinery. Lighting that is right for graphics can produce unflattering portraiture.

Lighting approaches
We can approach lighting in several ways. Each can produce successful pictures.

* We can position subject and camera *to suit available light.*
* We can *augment available light* to meet our purpose.
* We can *light from scratch*, relying entirely on our own specially-selected illuminators.

Unsuitable available light
Available light may not be appropriate for our purpose, or it may create unattractive effects.

Adjust the camera angle
Sometimes we can arrange our camera angle to suit prevailing light conditions.

Augmented light
We may be able to augment existing illumination to provide optimum results.

Special lighting
But on many occasions, selective lighting from a series of carefully positioned lamps will be necessary to produce effective pictures.

The Aims of Lighting

Whatever the subject, our lighting generally has similar fundamental purposes. We want to bring out characteristics that are appropriate for our needs, and perhaps suppress or conceal others. We light to give emphasis, or even to over-emphasise specific features. In broad terms, we usually light for visibility, for clarity, and for decorative effect.

Visibility

We could point a bright light or two at the subject from the camera position and it would be illuminated simply and economically! Why will such lighting not suffice? It may! But it will not usually produce an *effective* picture, to attract the viewer's eye. Pictures should persuade the viewer, and hold his attention. Visibility alone is seldom enough.

Clarity

Close scrutiny of the TV screen reveals how very approximate is this image to the real thing. Detail is limited, tonal values are restricted. And yet, in this picture we have to convey an illusion of the real full-colour, three-dimensional world about us. This is where methodical lighting can help us.

Our lighting is arranged to conjure up an illusion of solidity, to enable the viewer to create a mental image of the situation in front of the camera. If we rely on accidental lighting, then the very part of the subject we want to see may be in shadow; its shape or texture may be unclear. Communication fails. Information is lost.

Pictorial defects

If the direction of lighting is wrong for a particular application, we shall see pictorial defects of various kinds. Some of these are technical (page 24), and arise from the TV camera's behaviour under these conditions. But there are many more obvious, yet common, shortcomings. A lecturer's shadow falls over a map to which he is pointing. A bright reflection prevents our seeing part of an oil-painting. A demonstrator describes an article's rough surface that looks quite smooth in our picture. Careful lighting avoids such distractions and ambiguities.

Decorative lighting

Our lighting may be used decoratively. It can create pleasing pictorial effects that charm the eye and encourage interest. Visual appeal is an essential. Even where our audience is 'captive' (e.g. in a classroom), we must ensure that the picture not only conveys its message, but is enjoyable to watch.

Visibility
It is not sufficient just to illuminate the subject so that we can see it.

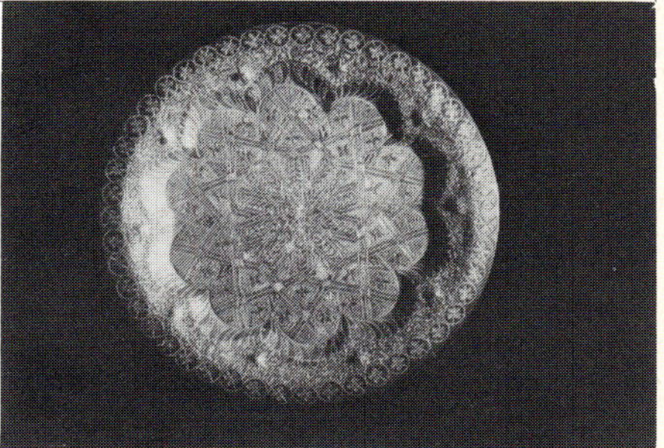

Clarity
If lighting is inappropriately angled, we may find that important subject detail is unclear in the picture.

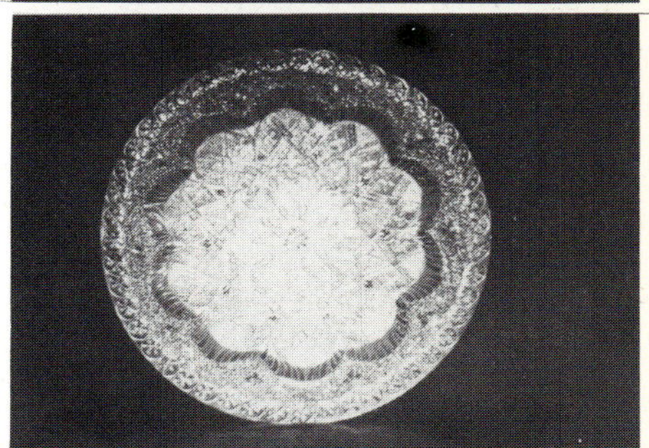

Effectiveness
Inappropriate lighting can actually prevent the audience from being able to see the information we want them to see!

Light has a deceptively variable quality.

The Character of Light

We usually accept light as we find it, without a second thought. Eye and brain adjust so that we are not even aware of its changing nature. The camera, however, makes no allowances for what the user is particularly interested in, but impartially records the image through its lens. So, if the lighting has certain characteristics, the camera tends to reveal these, even when the situation looks quite 'normal' to the person handling the camera.

The fundamental aspects

As we examine lighting techniques, it soon becomes apparent that we have several quite distinct factors to take into account.

The *quality* of the light (its dispersion) ranges from *hard* to *soft*. The *hard light* is very directional and casts well-defined shadows. This type of light is emitted by the sun, spotlights, and other concentrated sources (page 30). Conversely, *soft light* has a diffused, shadowless character. It is the scattered illumination that spreads from broad light sources, and from an overcast sky (page 32).

The *direction* of the light is all-important, for ultimately it determines how the subject and scene are modelled, and consequently, what they look like on camera from a particular viewpoint. It is not surprising, therefore, that light direction becomes one of our primary concerns in this study.

The *colour temperature* or 'colour quality' of the light is critical when we are shooting in colour (page 18). What appears to be 'white light' actually varies considerably, and may have a reddish, yellowish, or bluish quality. Colour quality differs widely with the type of light sources. It may also be coloured by reflection from nearby surfaces. Daylight may vary considerably and artificial light can be produced in a vast range of colour qualities.

The influence of light

Light influences where we look, how we feel about the scene before us, and determines how we interpret pictures. Light reveals and conceals. By our choice of lighting we can show texture, emphasise or suppress it. Our impression of shapes, spatial relationships, and size, are all modified by lighting treatment.

Little wonder, then, that if we take 'pot-luck' with the lighting conditions, accept inappropriate existing illumination, or misapply light, we cannot really hope to achieve the results we want.

Light quality
Hard light casts shadows, and can produce strong modelling of contours and texture in subjects. Soft light provides diffused, shadowless illumination. Under such lighting modelling and texture are less apparent.

The influence of light
The appearance of quite simple objects can be changed considerably by varying the lighting treatment.

The Colour Quality of Light

For *'black-and-white' systems*, colouration of the incident light matters only when this is strong enough to give noticeably inaccurate tonal reproduction. But any *colour system* is balanced to suit white light of a particular chromatic quality – i.e. with certain proportions of red, green and blue. So for colour accuracy the incident light's colour temperature should match the system's colour balance; although you may sometimes accept (even prefer!) inaccuracy – even introducing variations deliberately for effect.

Appropriate colour quality

Colour film is designed to suit light of a specified *colour temperature* (daylight, tungsten-halogen lighting, or photoflood light). If you use tungsten-balanced film in daylight, a blue colour-bias swamps the picture, while a daylight-balanced film produces an overall orange-yellow bias under tungsten lighting. However you can match for optimum colour balance by using appropriate colour filters on the camera lens (for overall correction) or over individual light sources (for separate individual correction).

The *TV camera*, on the other hand, is readily adjusted to suit the prevailing light by altering the respective red, green, blue channels of the camera system. So whether you are shooting under a clear blue sky (high colour temperature), or the yellowish-red light of household tungsten lamps, the camera can be instantly re-adjusted to match.

However, no system can cope with *mixed* luminants of different colour temperatures (e.g. daylight and tungsten light). Although the eye of anyone there might adapt to these differences and accept them, the camera shows distinct colouration for areas lit by unmatched sources. Similarly, while the eye accepts a dimmed tungsten lamp as simply 'less bright' than before, our camera detects the change of colour temperature resulting from dimming, and subjects are reproduced with a reddish-orange bias.

Correcting the light

You can introduce a *colour correcting filter* over a light source to convert its colour temperature to a higher or lower value. An amber-orange type (e.g. Wratten 85 series) lowers its colour temperature, while a blue filter (e.g. Wratten 80 series) raises it. *Colour compensating (cc) filters* of primary or secondary hues (red, green, blue, cyan, magenta, yellow) can correct particular deficiencies in the colour quality of the light source. A filter is usually rated in *mired units* (one million divided by its value in kelvins). To change a source to another colour temperature, you check the difference needed, convert this into *mireds* and use a filter of this value.

THE COLOUR OF LIGHT

Appropriate balance
For colour TV the camera system should be electronically balanced to suit the colour quality of the illumination being used.

Mixed luminants
Where the illumination is of mixed colour temperatures, the camera cannot provide a satisfactory match overall.

Compensatory light filtering
Instead, filter the light to suit one type of luminant (tungsten or daylight), and colour balance the camera channel to that standard.

Typical Light Sources

Various luminants are used in television lighting. Each has its particular advantages and limitations relative to cost, efficiency, light output, life, size, colour quality.

Fluorescent lighting
Banks of fluorescent lamps provide relatively diffuse illumination, but each tube has a comparatively low output, and scatters light over the scene. Their colour quality is not strictly suitable for colour systems.

Tungsten lamps
Low-power tungsten lamps (below 150 W) have only limited applications in TV lighting – chiefly for 'practical' decorative sources. Studio tungsten lamps (150 to 10 000 W) served film and TV monochrome studios for many years; but their light output and *colour temperature* fall in use, making them less suitable for colour systems.

Tungsten-halogen lamps/quartz lamps
Improved design (halogen gas filling) provides more compact, high-efficiency tungsten lamps with nearly constant output and colour quality throughout a longer life. Made with 'hard glass' envelopes (cheaper, shorter-lived) and quartz bulbs (handling discolours and damages their surface). Quartz lamps are used in point source and linear (strip) form to suit all types of fittings.

Internal-reflector/sealed-beam/PAR lights
Made in tungsten (incandescent) and tungsten-halogen forms, with regular or overrun versions. These all include internal silvered reflector surfaces, so need only a protective housing (no additional mirror or lens systems). Both pre-focused and diffuse designs provide lightweight adaptable sources.

Overrun lamps
These tungsten or quartz lamps give a much greater output for their wattage than normal, and a higher colour temperature – although with a much reduced life (e.g. 2 to 100 hours). Valuable for lightweight portable fittings, but liable to break down unexpectedly.

Metal halide lamps
In HMI, CSI, and CID forms, these very efficient enclosed arc lamps have a considerable light output at high colour temperature (around 5500 K), matching well with daylight. However, a bulky ballast unit is necessary to regulate the arc's current.

TYPICAL LIGHT SOURCES

Fluorescent lighting
Economical but bulky. Limited output. Dubious colour temperature. Large area illumination, not easily localised.

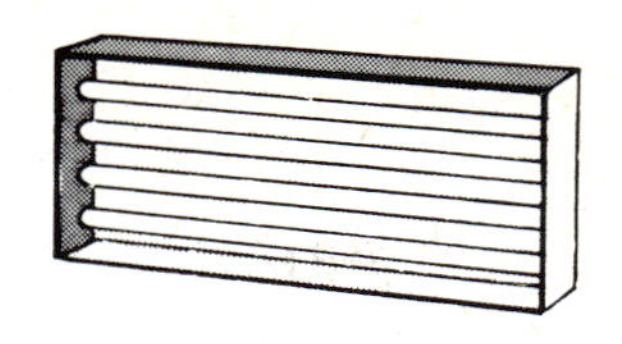

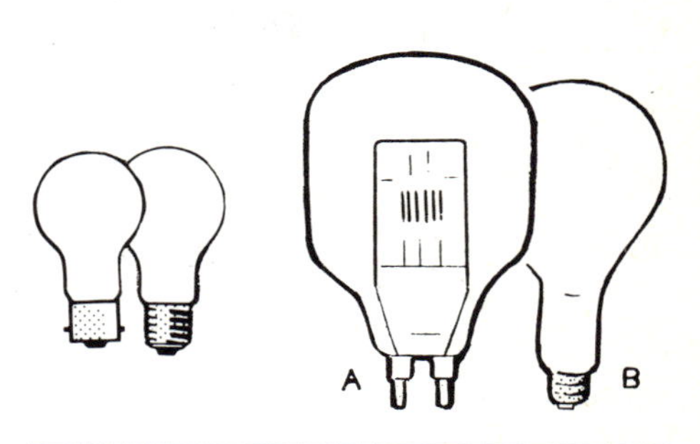

Tungsten lamps
Relatively cheap, but output and colour quality deteriorate badly in use. A. Fresnel spotlight. B. Scoop.

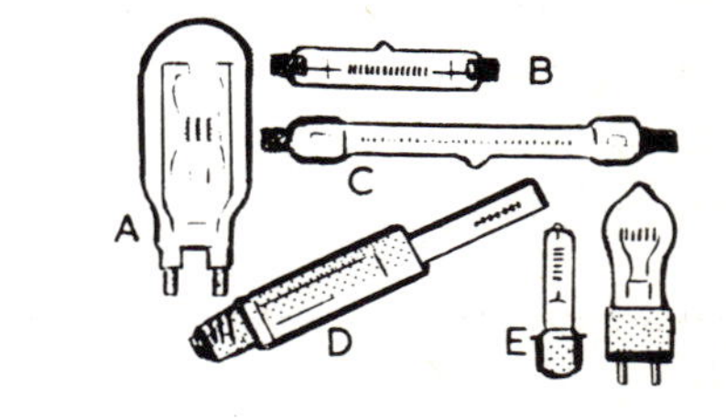

Tungsten-halogen lamps/quartz lamps
Compact adaptable light sources of constant output and colour temperature. Extremely hot and brittle in use. A. Fresnel spotlight. B. Lensless spot. C. Cyc light. D. Scoop. E. Effects projectors/ ellipsoidal spots.

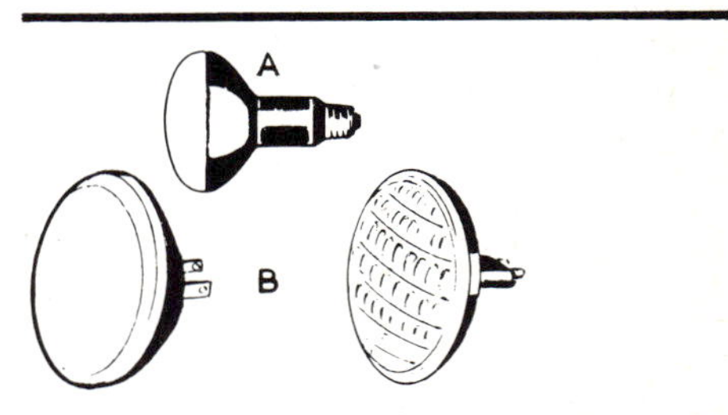

Internal-reflector lamps (A)/PAR lamps (B)
The lamp's integral silvered reflector directs and focuses the light. Available in spot and flood versions. No adjustment possible, but clip-on accessories control the light beam.

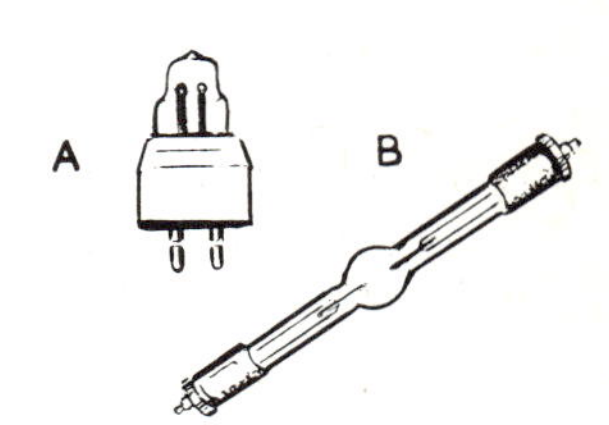

Metal-halide lamps/CSI (A)/HMI (B)
High effeciency, high output sources requiring bulky ignitor/ballast units. Take time to reach full intensity. May not permit switch-on while hot. Need a.c. supply.

Each type of light fitting has its own characteristic behaviour.

Types of Light Fitting

We can, for all practical purposes, regard lighting equipment as falling into three categories. Each has its own role in lighting techniques, although in advanced methods you will find these demarcations less rigid.

Soft-light sources

Ideally, these units produce diffuse, shadowless illumination. For really effective diffusion, however, light sources would have to be inconveniently large, so in practice one can often detect faint shadows, owing to the technical problem of designing compact, efficient 'soft lights'.

Soft light is used to provide overall *base light*, for *fill-light*, and for shadowless applications.

Soft-light fittings take several forms, each with its own advantages. Simplest is the *open reflector* with its single frosted bulb. These units (scoops, sky-pans, dish reflectors, broads) are mobile and adaptable, and most successful when grouped together. *Multi-lamp* designs have proved adaptable for TV purposes. A diffusing medium made of spun glass, wire mesh, or frosted glass, can help to soften the light further, and reduce multiple shadows. Typical units include four- and five-light strips, banks, 10-lites, clusters, nests, mini-brute. *Internally-reflected* light sources are also used in soft-light fittings.

Hard-light sources

These units provide hard, well-defined shadows; for creating modelling, revealing texture, for localised lighting. Hard light comes from small, concentrated point sources.

The hard light from a bare bulb can be used more efficiently if collected by a concave mirror and focused by a lens system. This is the arrangement of the familiar *Fresnel lens* spotlight, which is the mainstay of TV lighting. It may have a hard or a softened beam edge, the latter being more adaptable (page 88).

Light coverage by the Fresnel spotlight can usually be adjusted within limits, by *flooding* and *spotting* the fitting. However, this simultaneously affects its light intensity, and sometimes the evenness and hardness of the light beam (page 44).

Most Fresnel spotlights use tungsten or tungsten-halogen lamps. Some designs employ carbon arcs, discharge lamps, xenon and compact sources, but these have various limitations for the general user.

Projection spotlights (profile spots)

These are designed to project precisely-shaped (usually hard-edged) light beams, shadow patterns from metal stencils, or images from glass slides. A special optical projector unit can also project moving images derived from its large internal glass disc.

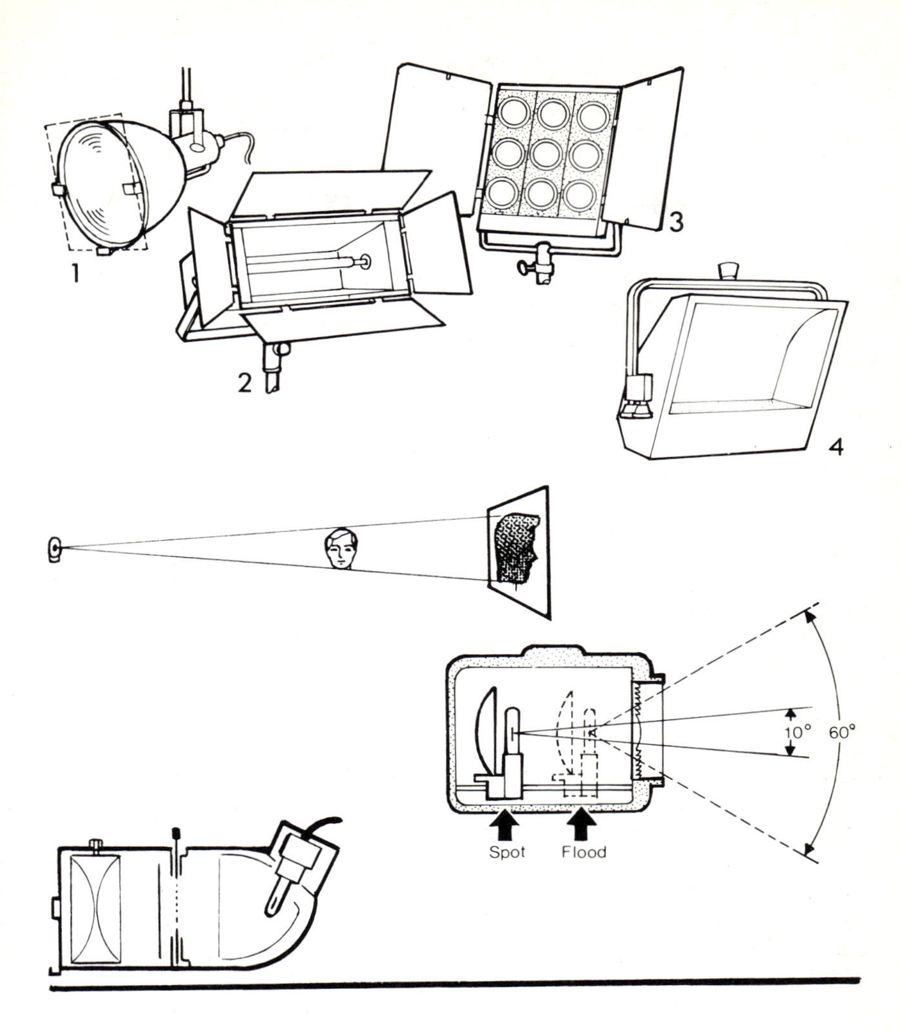

LIGHT FITTINGS

Soft-light sources
Typical sources for diffused light use large reflectors, line filaments, multiple lamps, or internally-reflected designs. 1. Scoop. 2. Small broad. 3. Floodlight bank. 4. Large broad.

Fresnel spotlight
Any *point source* produces hard light. In a *Fresnel spotlight* this light is projected as an adjustable soft-edged beam.

Projection spotlight/ellipsoidal spotlight/profile spot
Has a hard-edged light beam of adjustable shape. Projects patterns and shadow effects using a metal mask (gobo, cookie).

How Much Light is Required?

The average amount of light our camera tube requires to work efficiently, is determined by its inherent design (page 128), the lens aperture at which it is operated, and to some extent, by its electronic adjustments (target, beam, and video-gain control).

Apart from the *crushing-out* effects *(blooming)* of highlights and over-bright surfaces (pages 10, 12), excessive light can cause streaking. Insufficient light, and the picture is broken up with *video noise,* and an ectoplasmic smearing *(lag, trailing)* follows as images are moved across the frame.

Lens aperture

The light intensities *(light levels)* we need, and the lens *aperture (stop)* we select are directly interrelated. While larger apertures pass more light, so that lower illumination levels suffice, they result in a limited *depth of field. Stopping down* to a smaller aperture improves focused depth (especially in close shots), but necessitates proportionally more light. As a reduction of one full stop (e.g. *f*4 to *f*5.6) requires double the original light level on the scene, clearly lens apertures are not to be casually chosen!

In television camera work, we can see results immediately, and check any corrections quite readily.

Lighting intensities

If the picture is too light overall, we can either reduce corresponding light intensities, or stop down the lens to a smaller working aperture. Too dark overall, and we can either increase the light or *open up* the lens aperture. How do we judge, then, whether we have an exposure or a lighting problem? First check the prevailing light levels, and see whether these are reasonably normal for your equipment and working aperture (page 128). (There is usually a local upper-intensity limit, due to lamp and power availability, ventilation restrictions, etc.) If lighting is satisfactory, then correction is done by lens aperture (or target voltage adjustment in the vidicon).

When subjects appear *locally* over- or under-lit, the remedy is usually obvious. We reduce the brightness of a lamp that burns out the face of someone lit by it. We intensify a lamp where the subject is insufficiently bright (pages 46, 48, 124, 126). But appearances can be deceptive. Where, for instance, a disproportionate amount of light falls upon a background, the foreground subject appears by comparison to be under-lit. We must avoid the temptation of stopping down to accommodate the excess, but instead, expose for the correctly-lit foreground subject, and reduce the over-lit area. It is all too easy for lighting to escalate as we increase one source's output to match another that is too bright.

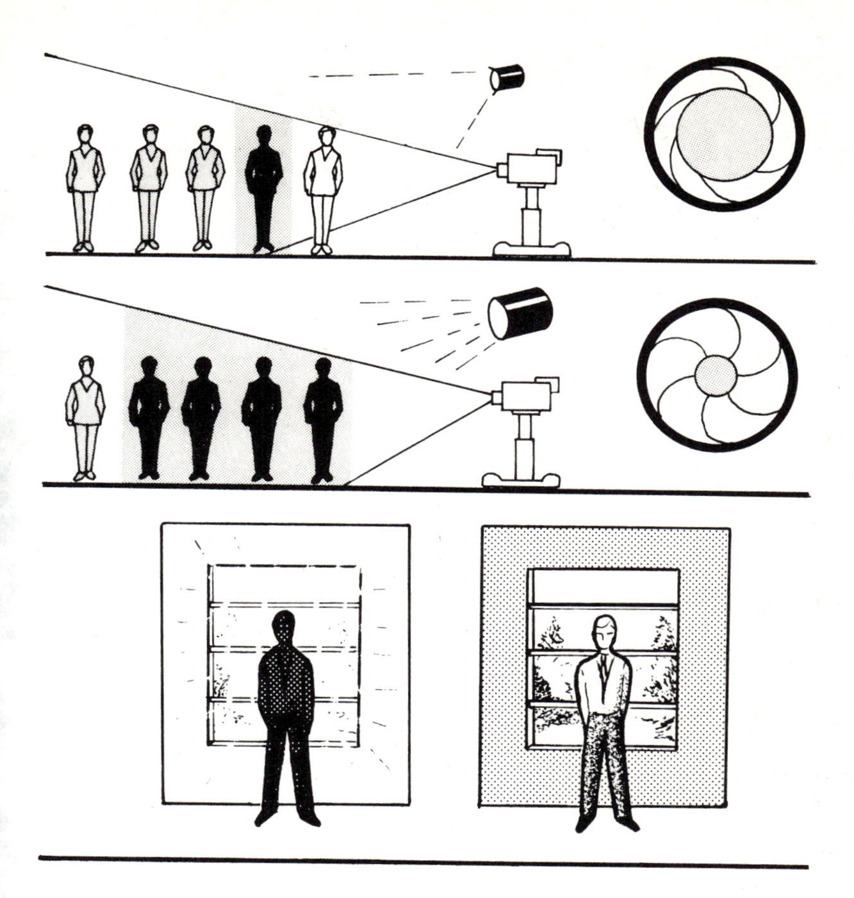

THE AMOUNT OF LIGHT REQUIRED

Lens aperture (stops)
At large apertures (e.g. *f*2) lower light levels are acceptable, but depth of field is shallow. At small apertures (e.g. *f*16) higher light intensities are necessary, but depth of field is considerably greater.

Relative exposures

$$\frac{(\text{second } f \text{ number})^2}{(\text{first } f \text{ number})^2} = \text{change in light levels required.}$$

Thus, stopping down from *f*4 to *f*8 needs light levels to be increased fourfold to maintain a similar exposure.

$$\frac{8^2}{4^2} = \frac{64}{16} = \frac{4}{1} \text{ light required}$$

Relative brightness
Against an overbright background, the subject appears insufficiently lit. Open the lens aperture to expose the subject correctly and reduce the background brightness.

On a tight budget, every lamp counts.

Is Much Equipment Needed?

The lighting equipment we need depends on the occasion. Portable gear must be lightweight, compact, yet robust, using the lowest possible power, quickly installed and dismantled, every lamp being used to maximum effect. Studios generally have a more generous allotment of heavy-duty lamps, ample power-availability, switching and dimming facilities. The area to be lit also influences the choice of lamps. More restricted action may require only a few lower-powered fittings, while a larger area needs higher-power lamps or a number of medium-power units. As lighting treatment becomes more elaborate the equipment needed quickly grows.

Simplicity or elaboration?

While some situations require only a single lamp for optimum effect, others depend on a skilful blend of many carefully set lamps. However, it is just as easy to over-elaborate the treatment, and create a fragmented fussy overall effect, as it is to use too few lamps, and miss attractive visual opportunities.

Lighting arrangements are often influenced by equipment shortages, limited power supplies, or just insufficient time. Sometimes a more sophisticated approach might not work where performers' positions vary unpredictably, and instead a simpler set-up would prove more reliable.

Practical considerations

You can achieve economies by such simple expedients as *supplemented daylight* (using auxiliary or reflected light), by using *non-restricted light* (e.g. a person's key light illuminates all the background also), or by using *dual-function lights* (e.g. one person's key light is his neighbour's back-light).

While simplicity is preferable to an over complexity, in which a myriad of lamps illuminate fragments of the scene, it brings its problems, too. If we use a lamp for dual functions, it could prove too bright for one purpose, and too dim for the other.

If a single lamp covers an appreciable area, we may find – particularly with soft-light sources – that it is too intense when the performer is close to it, and insufficiently bright for a distant position. Clearly, two separate lamps would be preferable, each lighting its own respective area properly.

If, on the other hand, we use several lamps to illuminate a particular area, we may have the difficulties of avoiding multiple shadows, conflicting light directions, and the modelling of one light nullifying that of another.

There is always the likelihood, too, that performers may move from their intended lighting, into that arranged for something else (e.g. from their own key light, to the beam of a bright lamp modelling dark drapes nearby).

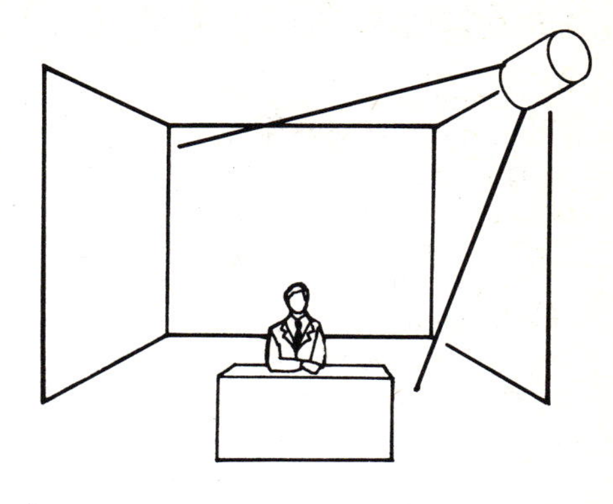

Basic lighting
A. Very basic illumination is liable to look crudely unattractive.

Elaborate lighting
B. Elaboration in lighting treatment may provide flexibility of effect – but may equally well result in confusing multiplicity.

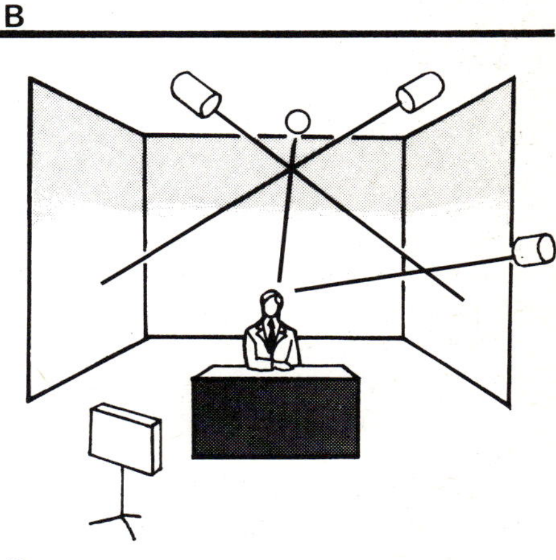

Balanced lighting
C. A well-balanced lighting treatment is both adaptable, and direct in its functions.

How Are Lamps Supported?

At the lightweight end of the scale, we have lamps that are quite powerful, yet small enough to be hand-held, or mounted on the camera. These are particularly useful where mobility is essential. For example, the *camera light (headlamp, basher)* is a utility light near the lens, which serves variously as frontal fill-light, to illuminate captions, or to supplement other illumination.

Slung lamps

Most smaller TV studios are fitted with a lighting *pipe-grid* just below a ceiling typically 3.5-m/12-ft high. Lamps are clamped, clipped or slung from this lattice structure of tubular steel. Power supplies for these lamps are usually arranged from ceiling points, or distributed adjacent to rigging positions.

Larger studios often use a series of independently-suspended bars (*barrels*), or a slotted ceiling grid into which telescopic lamp-hangers fit. *Catwalks* (walkways) may be provided around and above the studio, giving access to equipment and lamp positions for manned lights (e.g. spotlights).

Stand lamps

Lamps can be fixed into floor-stands of adjustable height. These enable us to position light with considerable precision. Unlike slung lamps, which are necessarily more cumbersome to re-locate, stand lamps are readily altered to suit the subject.

Unfortunately, floor-stands can be frustrating obstructions in highly mobile production, their cables requiring careful anticipatory routing. In many TV studios, therefore, such stand fittings are mainly utilised outside the actual operational areas (i.e. at the edges of settings, as 'sunlight' through windows, for graphics).

Clamped lamps

These are smaller lamps (e.g. 100 to 1000 W) held in fitments attached to the top, side, or face of scenic flats. Clamped lamps are valuable for localised backlight or keys (page 100), and to illuminate less accessible areas.

Ground lamps

Lamps are often laid on the ground, or fixed into very low stands, hidden behind scenery for upward lighting.

Provisional supports

On location, certain fit-up facilities extend lighting opportunities, including: lamp base-plates that can be stuck on almost anywhere with adhesive *gaffer tape*, telescopic lamp-support bars (wedging between floor and ceiling), and spring-jaw clamps (page 110).

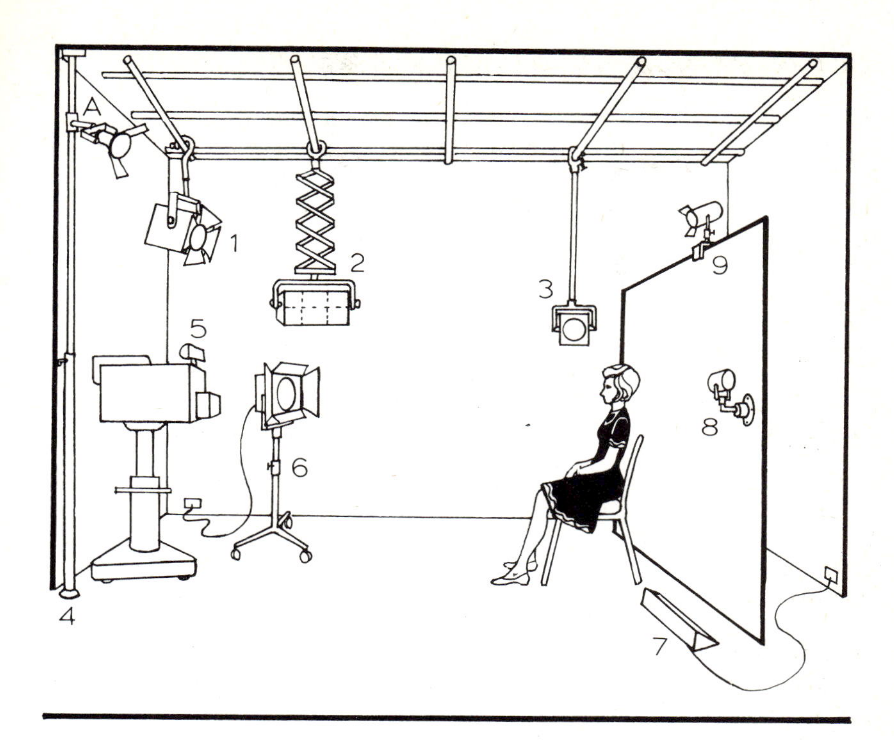

LAMP SUPPORTS

This assembly of typical lamp supports shows:

1. Spotlight clamped to 1¼-in *pipe-grid* (4-ft and 8-ft centres).
2. Soft light on *pantograph* (expanding, spring-balanced hanger).
3. Spotlight on *drop-arm telescope* (tubular support, fixed or adjustable).
4. Expanding pole (*polecat, barracuda*) clamped between ceiling and floor. Lamp attached at A.
5. *Camera light (headlamp, spot bar).*
6. Sectional telescopic *floor stand* (swivel castors or rigid feet).
7. *Ground row (trough).*
8. Spotlight in *face clamp*.
9. Spotlight in *flat clamp*.

Hard light produces modelling and shadows — deliberate and accidental.

A Closer Look at Hard Light

The skilled application of *hard light* is undoubtedly at the heart of most good lighting practice. It gives a picture of vigour and definition. Many subjects appear disappointingly dull and uninteresting if lit by soft light alone.

The merits of hard light

Hard light creates sharp, clear-cut modelling. It can reveal shape and form distinctly, showing the contouring and surface features of subjects clearly and unambiguously. Texture is well defined.

Hard light produces shadows, and shadows have both a graphical and a pictorial value in picture-making. They can delineate planes that are hidden from our viewpoint, showing their outline, and so giving us greater spatial information about the scene before us.

Shadows can be attractively decorative, as when the tracery from a fretted screen spreads over the floor. Shadows can also reveal details of an environment, as sunlight shines through a window to be cast evocatively onto a nearby wall.

Hard light has the great advantage that it is easily controlled. We can obstruct and shape the light beam by placing surfaces to restrict it (page 50). The light rays can be made parallel, and by lenses and mirrors manipulated for particular applications – e.g. reflected light effects, projected images.

The disadvantages of hard light

However, hard light has to be handled with a certain amount of skill, if we are to avoid seeing its pictorial weaknesses in our pictures.

Hard light can produce harsh, coarse picture quality of undesirably high contrast. 'Soot-and-whitewash' results are most likely when unsupplemented hard lighting is steeply angled, from above, or to the side of the subject. Used from the camera position, it results in unattractively flat, pasteboard images of our subject.

Every hard light produces its shadow. A six-lamp cluster can give rise to six adjacent shadows! A multi-lamp lighting set-up can result in a series of unrelated shadows that distract or confuse the viewer.

Shadows can be definitive. But, equally, their particular shape, size and clarity may be unsuitable for our purpose. For example, shadows may prevent our seeing surface detail, interrupt an otherwise plain surface, or conflict with our concepts of attractiveness or beauty.

Conclusions

Properly applied and controlled, hard light contributes dynamic, effective lighting. Usually it is blended with soft light, which enhances its pictorial impact.

USING HARD LIGHT

Revealing shadows
The shadows produced by hard light can reveal information that would otherwise not be apparent from the camera's viewpoint.

Evaluating shadows
Shadows may be pictorially effective or distracting, according to their appropriateness. Here an effect that is attractive in a long shot becomes unattractive in a closer view.

A Closer Look at Soft Light

Soft light is both a foil to the potential harshness of hard light treatment, and a form of lighting in its own right that is capable of producing great pictorial beauty. Without its use, certain subtle sensitive effects could not be achieved.

The merits of soft light
In TV, soft light is employed mainly as a *base light*, and as a *fill-light*.

Base light (foundation light) is an overall flood of soft light, ensuring that no part of the camera's shot remains unilluminated, and keeping tonal contrast to a minimum. At best, base light is a useful adjunct to *high-key* lighting, making all shadow detail clearly visible. At worst, it washes out surface modelling, and produces flat uninteresting pictures.

Fill-light is used specifically to illuminate the shadow areas cast by a hard light, to reveal information there, without causing extra shadows, or totally over-riding the impact of the hard light. The main light source (invariably a focused spotlight) is usually augmented by a subsidiary fill-light.

Soft light can be used as a *key light*, angled to the camera's lens axis, to provide delicately graduated half-tone shading. Located nearer the lens position, diffused light spreads over the subject, often making surface contouring and texture quite invisible. Only outline, and surface markings are clear. Whenever we are seeking to light an undulating surface to look flat or unmodelled, this control of surface modelling can prove useful.

The disadvantages of soft light
Soft light is not readily restricted. Consequently, in using it to light a subject, we are liable to find it spreading to adjacent areas. Even quite large screens placed around a soft source will not localise its light if this is truly diffuse.

Unlike focused hard light, the effective strength of soft light falls away rapidly as the distance from source to subject is increased. Consequently, a close subject may be over-lit, while another a little further away may be insufficiently illuminated by the same lamp.

Soft light all too easily results in flat, unmodelled, characterless pictures; especially if used from several directions. 'Overall lighting' does not provide optimum clarity, particularly in monochrome pictures.

Conclusions
Soft light should be used sparingly. There is a temptation to introduce it in order to avoid or suppress shadows. If those shadows are due to badly positioned hard light, the result seldom rises above the mediocre.

1

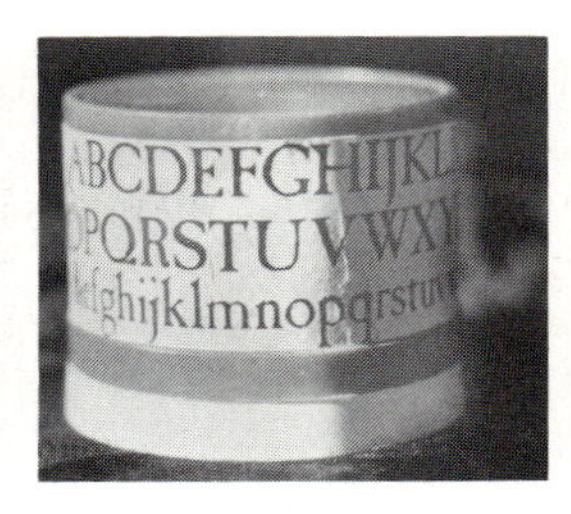

2

3

USING SOFT LIGHT

Base light
Diffuse overall illumination from an assembly of soft-light sources can provide a base light to reduce tonal contrast and ensure no under-exposed areas.

Fill-light
1. Lit by hard light alone, shadow areas may be too dense, and information obscured.
2. Appropriate fill-light reveals information without destroying form.
3. Excess fill-light and the modelling created by the key light is lost.

Light direction is always assessed relative to the camera's position.

The Effect of Light Direction

The exact direction from which light falls upon our subject has a considerable effect on its appearance. The lamp's position is all-important for predictable lighting. Fortunately, we can estimate the result of light orientation quite readily. There are three basic light directions, and any lamp position will produce certain of these characteristics.

Frontal light

We interpret 'surface contours' in a picture by the shadow formations, shading and tonal contrast that reveal their shape. Similarly, our impression of 'texture' comes from the tiny shadow variations that result from surface irregularities. If you light from a *dead-frontal* position along the lens axis, no shadows are visible, so signs of texture or surface undulations are minimal (page 120).

Frontal light is more usually placed some 10° to 50° off the lens axis, according to the extent to which we want to emphasise modelling and texture. As we move the key light further round the subject from its dead-frontal position, shadows of surface projections grow (a nose shadow spreads across the cheek, for example), and the subject planes furthest from the lamp begin to darken as they are left in increasing shade.

Side light (edge light)

When the light source is almost in line with a surface (*edge-lighting*), any irregularities are given maximum emphasis. Located to the sides of the subject, this effect is often termed *side light;* while directly above it, the direction becomes *top light*. Both are forms of edge light, but may emphasise different aspects of the subject.

Under edge-lighting, even slight undulations that are quite unseen with frontal lighting are thrown into sharp relief. Obliquely-angled light reveals surface features of such materials as paper, fabric, wood, stone, and emphasises detail in carved bas-relief, coins, etc.

Backlight

Moving our light source still further round behind the subject, frontal areas fall into shadow. (The far side of the subject is well lit, of course, as another camera in that position would show.) Now the only aspects we can see illuminated in a solid subject will normally be its top and side edges – in the case of a person, usually the top of his head and shoulders.

If we are lighting a translucent subject, the backlight will probably illuminate it more strongly than light from a more frontal position, and may reveal its structure.

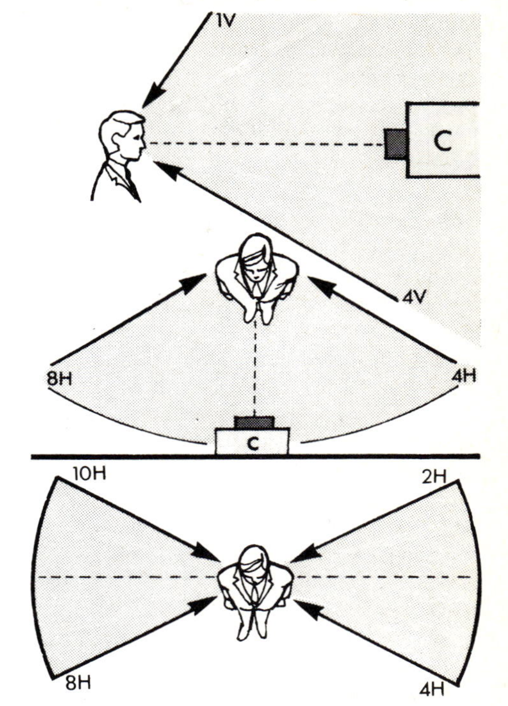

Frontal light
The main frontal light (key) normally constitutes the chief source of illumination, and largely determines the camera's exposure.

Side light
Side light usually emphasises contour and texture, throwing them into strong relief.

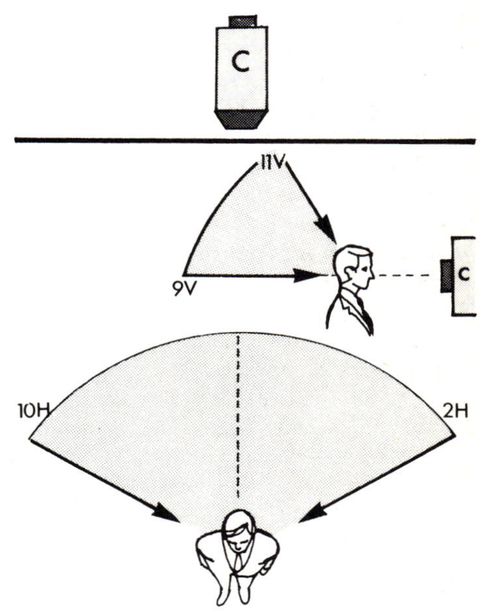

Backlight
Backlight rims the edges and top of subjects; shining through translucent or transparent planes.

With the parameters of three-point lighting in mind, our lighting treatment begins.

Where Do We Place the Lamps?

We have now met the principles that underlie all lighting techniques: the characteristic behaviour of *hard* and *soft light*, and the three extremes of *light direction*. The basis of lighting is to be able to apply these principles to the subject and its surroundings, so that they appear as we wish them to in the television picture.

Three-point lighting

The simplest complete lighting set-up is an arrangement often referred to as *three-point lighting*. We can apply this concept equally well for any three-dimensional subject. It is no ritualistic routine. We do not always use all three parameters. We may, for instance, use only a single key light, and nothing else. Our choice depends upon our purpose.

The *key light* is the strongest lamp; the main source of illumination. It usually establishes light direction, creating the principal modelling and shadow formation in the subject. Its most effective position will alter with the type of subject we are lighting, how it is positioned, the aspects we want to emphasise and any external considerations (such as the position of windows in the set). Never use two or more key lights. The result will be conflicting or nullified modelling, dual shadow formations, that detract from the overall effect.

The *fill-light (filler, fill-in)* is arranged next, to illuminate the shadow areas not lit by the key light. Not too strongly, for we do not want to overwhelm the illusion created by the key. Remember, the fill-light always has a supplementary role. It should never dominate. It is often possible to light very successfully without using any fill-light at all; but the results may prove to be rather more dramatic and contrasty than is appropriate.

Backlight is quite a bone of contention. Some people introduce it as a matter of course, but backlight is best applied very selectively, both in amount and direction. In illuminating the edge-contours of the subject, backlight often reveals their depth and form. Where both subject and background are of similar tones, backlight will also help to rim the subject outline so that we can distinguish between planes more easily. Without this treatment, subject and background may merge.

Designating direction

To indicate a lamp's position in space quickly, we need a simple, easily remembered, unambiguous system. The clock-face scheme outlined opposite has these advantages. By giving each lamp its horizontal and vertical references, we can locate a lamp's function for a particular viewpoint. In practice, people usually transcribe 'clock positions' more accurately than angular differences, but if necessary they can, of course, be converted instantly (5 minutes = 30°).

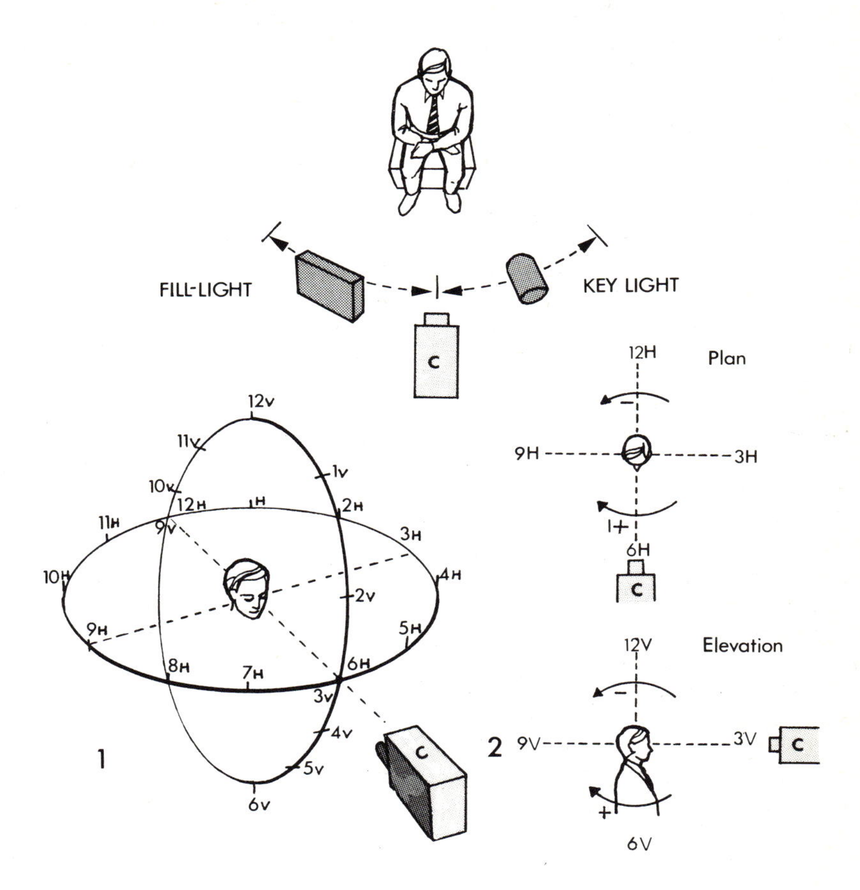

PLACING LAMPS

Three-point lighting
This basic three-light set-up (key, fill and back lights), known as 'three-point lighting', is used for most three-dimensional subjects.

Specifying lamp positions
1. The position of any lamp can be located relative to an imaginary horizontal clock-face with the subject at its centre, and a similarly centred vertical clock.
2. The camera position is at 6H/3V. Intermediate positions between 'hours' are shown by + (clockwise) or − (anticlockwise) signs. 'Hours' represent 30° steps; 'minutes' are 6° each.

Positioning the Key Light

For many subjects, we shall find that our key light can be located around 10° to 50° off the lens axis. This may provide appropriate clarity of detail and modelling for your purpose. But you will often need to locate the key more precisely than this, to avoid unattractive effects (spurious reflections, ugly shadows, distracting *hot-spots*), or to get exactly the visual impact you want. Whether this position is critical or uncritical depends both on the subject and the effect you are aiming for. The optimum angle and coverage may be a matter of careful judgment, and personal choice.

Choosing the optimum position
Fundamentally, the key light is usually the illumination that 'makes the subject visible'. But a well-chosen key does so much more than this. It largely determines *what the subject looks like*.

As we move the key light away from the lens axis, we shall see certain changes taking place. The resulting shadows move in the *opposite direction.* The further round we move the key, the more will these shadows be extended or displaced. Raising the key light, its associated shadows move downwards.

So, for example, as we steepen the lighting angle for a person wearing a cap, its peak shadow moves down, covering the wearer's eyes. Now we have to make a decision. Do we want eyes to be obscured by the shadow? (To hide their expression, perhaps, in very dramatic situations.) If not, we have the choice of retaining the overall strong modelling, but adding fill-light to illuminate the shadow (page 40). Or, if the shadow is intrusive, we may lower the key to shorten it (or, alternatively, tilt back or remove the cap!). Clearly, the position of the main lamp is important, if we want particular aspects of the subject to be distinctly and appropriately lit.

Where the most interesting features of the subject are not facing the camera (e.g. in profile) this will often guide us to the preferred direction for the key. Again, the actual key location relates to the aspects we most want to emphasise or display. The lamp could be in a frontal, side, or even a backlight position. Admittedly, the last would be for special types of subjects which were translucent, transparent, or had outline tracery, but this exception does point to the variations that are, in fact, possible when the situation arises.

KEY-LIGHT PLACING

Non-critical key direction
Some subjects are equally effectively presented using a variety of key directions; although appearance changes correspondingly with each different arrangement.

Critical key direction
Other subjects are ineffectively presented, or lose their identity if the key direction is inappropriate.

Positioning the Fill-Light

Fill-light position is one of those more contentious matters in lighting techniques. The professional tends to have his own argued preferences; although in the TV studios he has all too often in practice to rely largely upon slung soft-light units that are unsuitably steep for their purpose. The amateur tends either to ignore or to over-use fill-light.

The character of fill-light

Fill-light is primarily intended to relieve shadows, and render them 'transparent' (page 36). It should, therefore, be neither over-bright, nor itself cast shadows. So, ideally, fill-light should be soft.

There are exceptions. Where we have to introduce localised fill-light from a considerable distance away (e.g. for a singer lit by a spotlight on a darkened stage), the use of soft light is impracticable. Soft light does not 'travel well' (page 32) and it spreads. Then, as an expedient, we may have to resort to low intensity hard light to provide the fill-light, relying on the strong key light to swamp its subsidiary shadow. But this is not desirable as a regular practice.

Locating the fill-light

Some people like to place fill-light on the camera itself, as a camera light (*headlamp, basher, spot bar* or *frame*). As always, there are snags. The fill-light can dazzle the performer when working to the camera, reflect in his spectacles and background, and may raise difficulties when cameras move around during multi-angle shooting. The camera light has several advantages, though. It always relates to the camera's position. It is around lens height, and fills the shadows seen by the camera. The fill-light also serves to illuminate close subjects.

Floor-stand fill-light can be used to place soft light at a height of around 1 m/3 to 4 ft. Very effective for both seated and standing performers, this tends to compensate for the downward diagonal shadows cast by slung key lights. For many purposes, the lamp is located in the region of 6H to 8H, for a key coming from 4H to 6H respectively.

Slung fill-light has the mechanical advantage that it leaves the studio floor space clear for cameras to move about, and reduces the snaking floor cables to a minimum. Artistically, there is the likelihood that the fill-light is not exactly where we want it, that it will be too steep, and that it will spread over the entire setting. One can occasionally compensate for this relatively steep fill-light, by using a camera light or even an additional ground lamp under-lighting the subject.

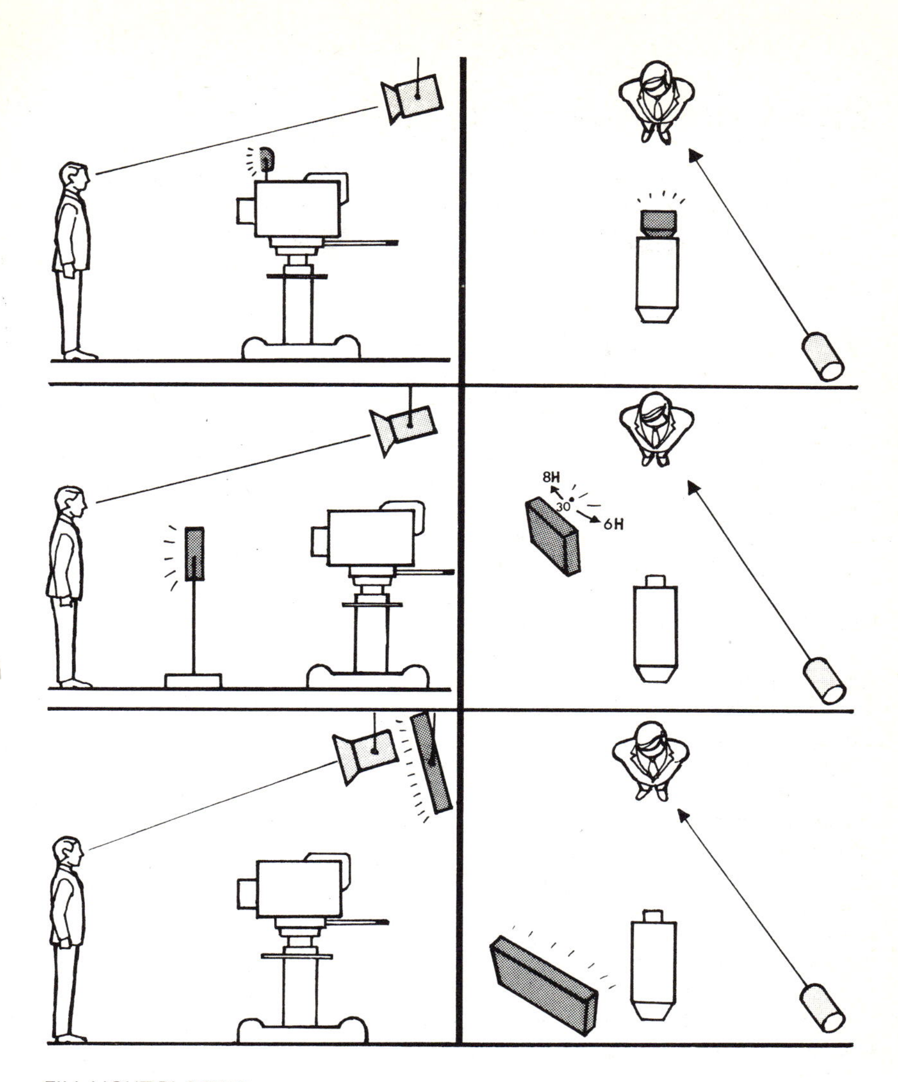

FILL LIGHT PLACING

Camera light
Fill-light may sometimes be provided by a small lightweight lamp attached to the camera. Its intensity may be remotely controlled, e.g. at the lighting panel.

Floor stand fill-light
Soft light affixed to a floor stand is easily adjusted to suit the subject, but may obstruct camera movements or come into shot.

Slung fill-light
Where cameras move around, so that floor lamps are intrusive, *slung* soft light has the advantage that it leaves the floor clear. However, its vertical angle may prove undesirably steep, and its coverage uncontrollable.

Positioning the Backlight

Backlight improves the attractiveness, clarity, and peripheral modelling of most subjects. Its position is not generally critical, although careful angling for individual subjects helps us to bring out particular features.

Single backlight
We can position a lamp some distance behind and above our subject, to cast a single forward shadow over the floor in front of it. This is an effective technique for large-area lighting, where the shadow becomes part of the decorative composition – as in dance, musical recitals, and similar displays.

We have to take care, though, to ensure that the shadow cast by backlight does not obscure detail in important parts of the scene. This problem arises most frequently in demonstrations (page 72), where shadows can fall upon items being discussed, so reducing their clarity. We must avoid the demonstrator having to work in his own shadow.

If we do not place our single backlight directly behind the subject, its forward shadow will spread diagonally over the foreground. This sometimes results in an unbalanced, lop-sided compositional effect, but it does have the advantage that the offset backlight rims part of the side edge of the subject. Dead backlight tends to light its top edges only.

Dual backlights
By using dual backlights, offset either side of the subject, we ensure that both side edges are clearly defined. The overall impression is of a rim of light round its periphery. This may be successful, but it should not be used as a routine. It can inadvertently 'pretty-up' or appear over-decorative. Dual backlights can produce a balanced pair of forward ornamental shadows, but these can become more confusing than a single lamp treatment.

Practical advice
Generally speaking, we shall find that the most used form of backlight employs a single lamp, slightly offset to the side of the subject opposite from the key light. The lamp should be high enough to prevent lens flares (the top barndoor can shade off the camera), or the lamp itself coming into shot. Restrict its vertical angle, however, for extremely steep backlight creates ugly effects (page 52). Top light has nothing to recommend it in most instances, for it over-lights and apparently flattens the horizontal planes of the subject.

BACKLIGHT PLACING

Single central backlight
A particularly effective treatment for decorative long shots. But its direction tends to leave the subject's side edges unlit.

Single offset backlight
Now the side edge nearer the lamp is illuminated, as well as the top surfaces of the subject. The resultant diagonal shadow from an offset backlight may look compositionally unbalanced in long shots.

Balanced dual backlights
Double-rim lighting has the advantage that it illuminates both side edges of the subject. But in longer shots the cross shadows may appear distracting.

How Do We Adjust Brightness?

The amount of light we want to fall on the subject and scene from different directions, will vary considerably with the effect we are seeking. We can adjust individual lamp brightness in various ways.

By adjusting power
The most obvious method of adjusting the light intensity is by choosing a lamp of appropriate power. We can use a lower power lamp relatively close to the subject, and obtain a similar light level to that given by a more powerful lamp further away. The close lamp will have a comparatively restricted coverage (and might intrude into the camera's shot). A distant lamp has a wider light spread (but could cast shadows of the camera and sound booms, etc., onto the scene).

By adjusting distance
We can sometimes alter the effective brightness of a subject sufficiently by making changes in lamp distance. This method is most effective with soft light sources on floor-stands, owing to the rapid fall-off characteristics of diffused lighting.

By adjusting spotlight focusing
As a Fresnel spotlight is focused (*spotted*) its coverage diminishes and its intensity increases. *Flooding* the spotlight provides a greater coverage, and the light intensity falls. Hence we often use the lamp's focus adjustments to vary its intensity.

By placing a diffuser over the lamp
When a diffuser (page 48) is placed over any lamp, it reduces its light output. How much, will depend upon the density of the material used. This is a particularly adaptable method, for we can modify the brightness of part of the light beam, leaving the remainder of it at full intensity.

By electrical means
As the voltage of the power supply to a lamp is reduced, we lower its light output. A *dimmer system* enables us to do this progressively, and to set a lamp at any selected intensity from maximum to 'black out' (i.e. extinguished), or to fade the lamp up or down (brighter or dimmer) for specific effects. Several types of dimmer are currently used, but the commonest include *SCR (thyristor)*, resistance, and *auto-transformer* systems. Unlike mechanical methods of lamp output control, electric dimming reduces the colour temperature of the light (page 18).

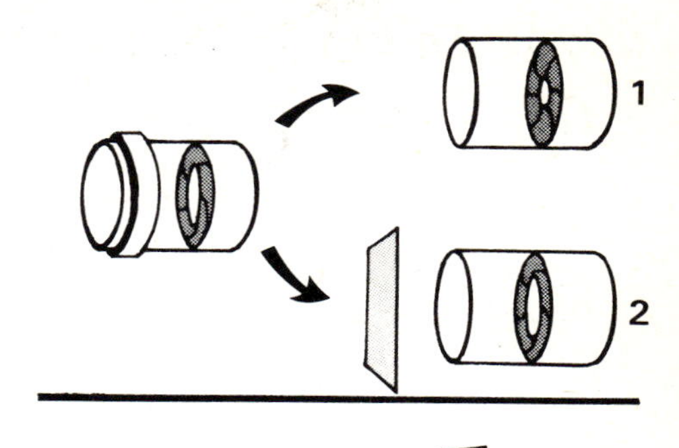

Adjusting overall image brightness
By stopping down the lens 1, or placing a neutral-density filter over the lens 2, we can reduce the overall light from the scene, and hence the exposure of the camera tube.

Adjusting individual lamps
To reduce the light level we can use a lower-power lamp at the same distance from the subject.

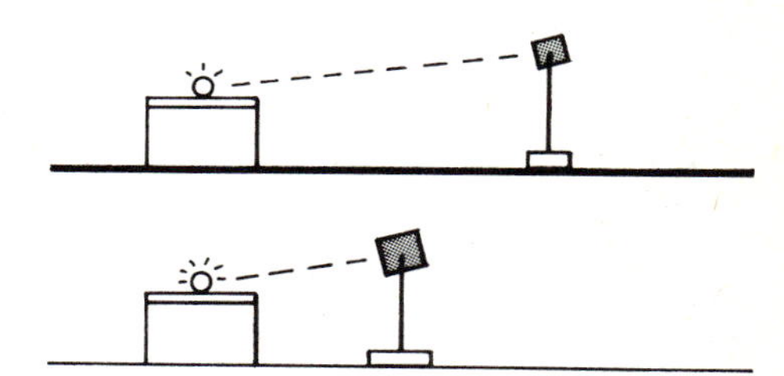

Increasing a lamp's distance from the subject reduces its effective intensity.

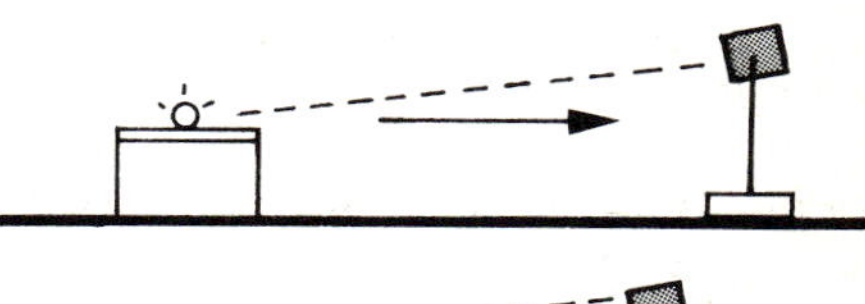

A Fresnel spotlight can be flooded to reduce its intensity.

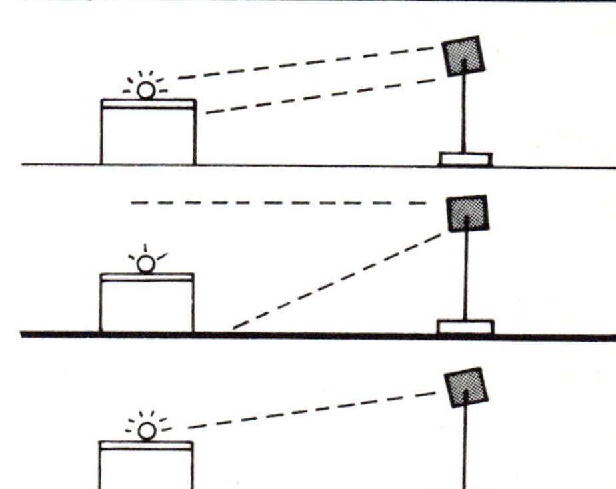

A diffuser positioned in front of a lamp will reduce its intensity.

What Is Lighting Balance?

When we adjust the relative intensities of our lamps to obtain a particular pictorial effect, we are controlling the *lighting balance*. Our aim may be simply to provide pleasing tonal relationships in a picture. But more subtly, the lighting balance can convey a mood, suggest a certain environment, or direct the viewer's eye to chosen parts of the scene.

Adjusting lighting balance

Lighting balance begins with our choice and arrangement of lamps; their relative power, distance, and diffusion (page 44). When setting the lamps, and subsequently watching rehearsal pictures, we see where the brightness of individual sources needs adjustment (*trimming*) and alter them accordingly to suit the effect we have in mind (pages 124, 126, 128). As a general guide, for most purposes, the key light will be strongest and the relative strengths of key-fill-backlights in a ratio of around 3 : 2 : 2 or 3 : 2 : 1 respectively. Faces are often made about 1½ to twice as bright as their backgrounds.

Using a base light

A lighting technique that originated with earlier types of camera tube, is that of using a fairly strong diffuse *base light (foundation light)* to illuminate the entire scene (page 32). All other lighting is then superimposed upon this, each lamp being made correspondingly brighter, according to the effect required.

This method has the merit of simplicity, and of avoiding undue contrast. However, its high-key, low-contrast results are not always appropriate, and can lack pictorial appeal. At worst, high intensity base light produces uninteresting illumination, in which modelling and texture are largely lost. For most purposes a base light as such is unnecessary.

Localised lighting

Another technique is widely used in TV and motion pictures. First we adjust the intensity of the key light to obtain the required exposure. Then fill-light is added, until shadow areas are sufficiently illuminated to reveal detail, but without losing the modelling created by the key. Backlight may then be increased until it outlines our subject; but not so strongly as to result in a distractingly over-bright rim. Background lighting (scenic, effects light) is then balanced to an appropriate level (pages 24, 88).

The resulting lighting can provide attractively dynamic pictures, achieving a three-dimensional illusion of form and space.

LIGHTING BALANCE

Changing the relative brightness of the various lamps lighting the subject and its surroundings can have a considerable influence upon their appearance and the prevailing mood.

Using Diffusers

Although all termed 'diffusers', these useful materials are introduced in practice for two quite distinct purposes: to diffuse illumination, and to control light intensity. Diffusers enable us to achieve effects that would otherwise require much more elaborate or extensive facilities, and to light with selective finesse.

Softening light

We can use diffusers to disperse light; to diffuse and soften its quality. For this purpose we employ translucent sheeting made from spun glass fabric, frosted plastic, or glass. These are often permanently fitted over the front of the light source – preferably a short distance away, to allow air circulation, and so prevent early lamp burn-out through overheating.

In practice, any so-called 'soft light' sources produce quite discernible shadows. Diffusers can help us here, to assist light-scatter and improve the degree of dispersal.

A diffuser placed over a Fresnel spotlight can enable us to 'soften off' this hard light source, and provide less defined shadows.

Reducing overall intensity

Diffusers can also be used, either in these forms or as wire mesh (open, or within a plastic sheeting), to cut down the overall light from any lamp that is too bright for a particular purpose (page 18). By altering the density of the diffusers (using a sandwich of one, two or three 'wires'), we can balance light levels with considerable accuracy, without resorting to electric dimmers.

In fact, by a careful choice of lamp power, and by the use of diffusers to reduce light intensities where necessary, a studio can present high-grade lighting treatment, even where it has no dimmer-board facilities whatever. Or again, where only a few *dimmers* are available, these can be used for lamps that must have continuous intensity control (e.g. for effects – pages 24, 116), while other lamps are adjusted instead by introducing diffusers.

Reducing localised intensity

We can use a small piece of diffuser, clipped into position, or held before a lamp, to reduce the light intensity falling on selected areas.

Sometimes part of a subject is light-toned, while nearby surroundings are dark. Then we may arrange for the full power of the lamp to illuminate the darker region and yet restrain it in another area with a carefully positioned diffuser, which prevents our over-lighting the light-toned subject.

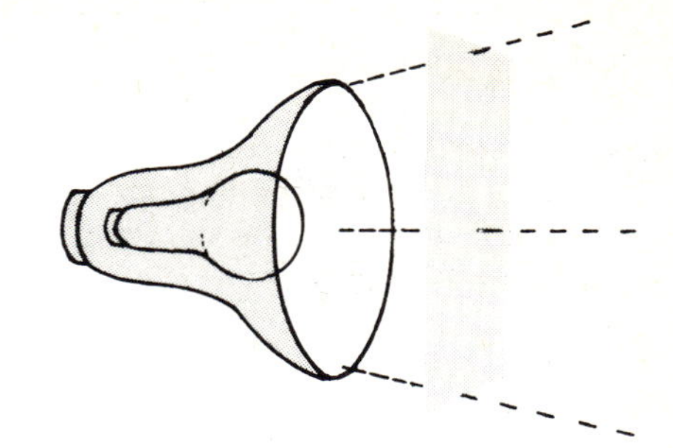

Softening light
A diffuser can help to soften the illumination from a soft-light source even further.

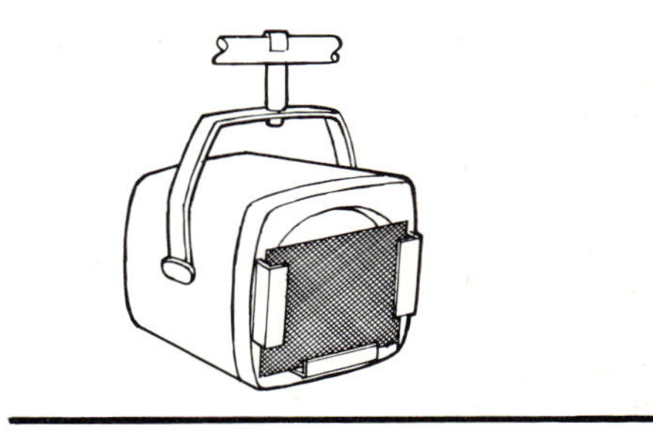

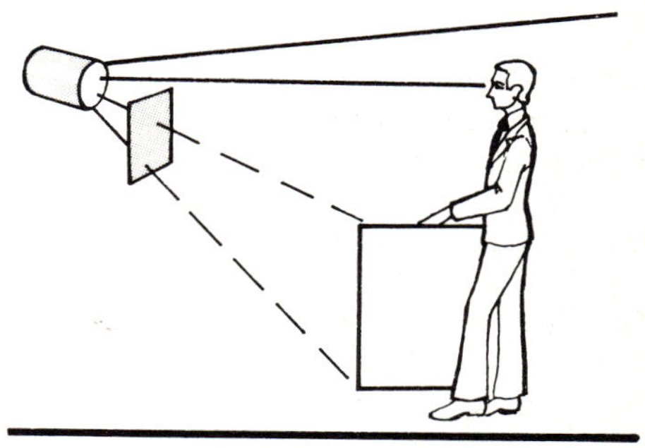

Reducing overall intensity
A diffuser can reduce light intensity from a lamp either to prevent it over-lighting a subject, or to balance its output relative to other lamps.

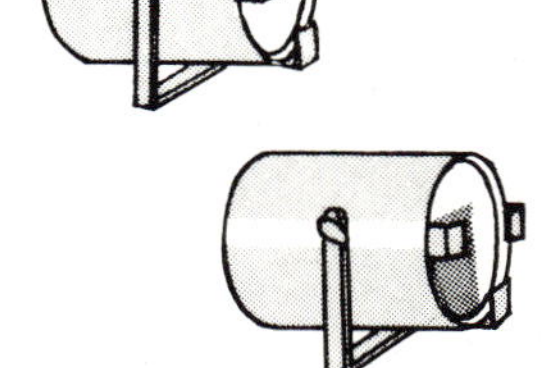

Reducing localised intensity
We can reduce the light falling on specific areas, by holding a piece of diffuser in an appropriate part of the light beam.

Restricting Light

It is seldom sufficient to allow illumination to spread uncontrolled over the scene. Each lamp usually has a specific purpose, and we want to confine it to this area alone. If light spills around, we are liable to see random streaks and shadows that distract the eye and spoil the picture. Several devices are regularly used to enable us to control the coverage of our lighting.

Barndoors

A metal frame is clipped to the front of a lamp housing, and fitted with two or four separately adjustable hinged metal flaps. The device can be rotated, to position the barndoor assembly at different angles.

Barndoors are extensively used to prevent the beam of one spotlight overlapping onto an area lit by another, or to restrict light to a localised region. Thus we can avoid over-lighting, prevent lamps from casting unwanted shadows, avoid multiple shadows, and similar quandaries.

Barndoors represent one of the most valuable lighting accessories. To the novice they give confidence, for he is able to contain his lighting to the place where he wants it. To the expert they provide the means for precisely localising light as he builds up a systematic treatment.

Gobos, flags, cookies

You can use *gobos* (sheets of wood, metal or cloth) to prevent light spilling over large areas, or to hide a lamp that is in shot. By placing a *flag* (small gobo) in front of a lamp, you can cut off light from a selected area more sharply than a barndoor would allow.

A *cookie* is a metal sheet stencilled or fretted into irregular shapes. You can place it in front of a spotlight to cast dappled light over a surface and provide a patterned or broken-up effect.

Spill rings

These are made up from a series of shallow concentric cylinders. They help to reduce the light-scatter from certain types of fitting, including scoops, internal-reflector (sealed-beam) lamps, and other open-fronted lamp housings.

Snoots

These cylindrical or conical tubes are sometimes clipped to the front of a lamp housing to restrict the spread of its light to a small circular region. The 'spotlight' effect can help us to pick out a detail of a subject, or to highlight a specific area.

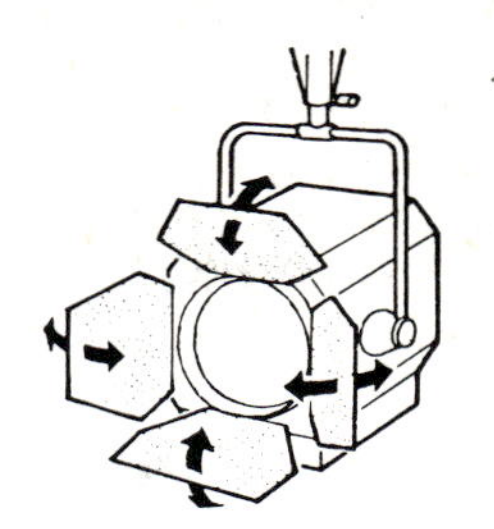

Barndoors
Barndoors are designed to provide selective cut-off of the light beam. Door flaps can be adjusted individually or in combination. Both two- and four-door designs are used.

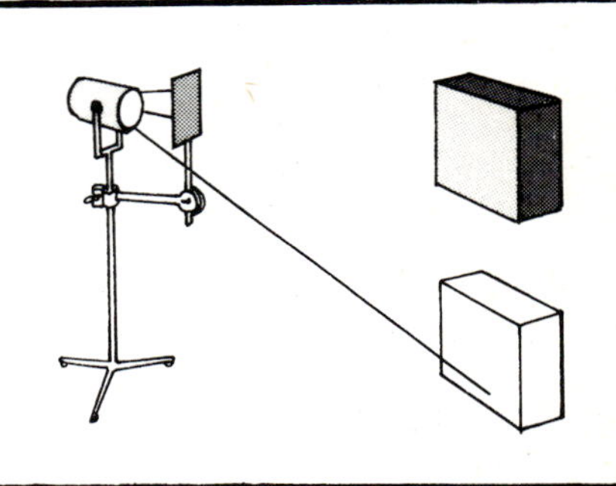

Flag
The flag supported in front of a lamp (by a flag-holder, or flag stand), casts a shadow, to keep light off a particular region. Metal and wooden sheets are used.

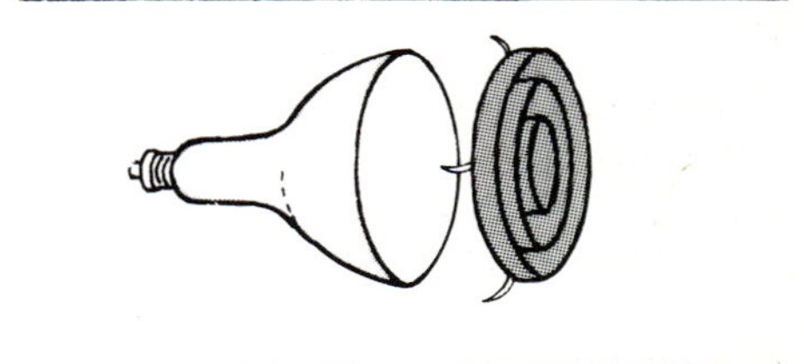

Spill rings
Spill rings direct light forwards, reducing the sideways spread of illumination

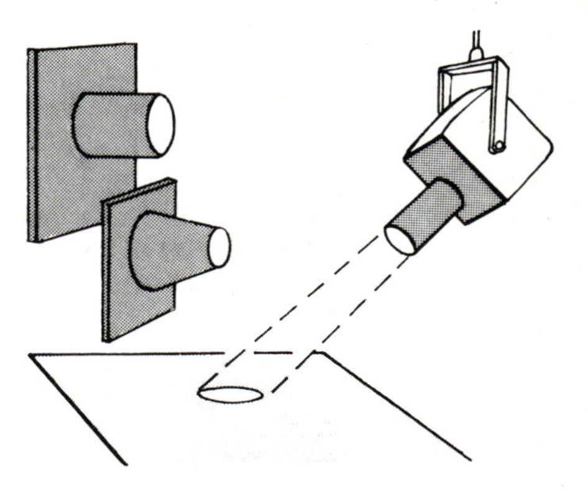

Snoot
A metal snoot confines the light beam to a small circular area. Snoots in a variety of shapes are used.

Beauty is certainly in the eye of the beholder!

Portraiture

Most of us tend to be pretty critical when assessing pictures of people. Other subjects can be lit in a variety of fashions without arousing comment, but not the human face. As people constitute our main TV subject, this means that careful portraiture techniques are essential.

Although one may occasionally, for dramatic purposes, light people to make them appear grotesque, aged, etc., this is exceptional. For most purposes, such treatment would be quite out of place. The object of the exercise is to avoid getting such results *accidentally*!

Lighting women

By 'good portrait lighting' we usually mean a flattering, attractive result that minimises the worst aspects, and maximises the best.

With portraits of *women*, it is generally best to avoid vertical or edge lighting of any kind. If a key light is steep, eye-sockets can become black recesses, bags under the eyes become prominent, shadows grow to exaggerate wrinkles and any other facial irregularities.

Instead, lighting should be reasonably frontal. Softened-off hard light, or diffused lighting can be very effective. A woman's backlight can be a little stronger than one might use for a man, to give definition to her hair. Double-rim backlighting can add a touch of glamour to the portrait. A camera-light can be used to give life to the eyes, with small single *catchlights (eyelights)*.

If we want to be particularly flattering, a low-power lamp can be used from just below waist level, to light upwards and nullify shadows and modelling being cast by the normal lighting set-up, However, this *under-lighting* can be overdone. Like *dead-frontal* lighting or excessive soft light, it is liable to flatten out the face so much that it loses character and becomes 'pudding-like'. Nor should we automatically apply such compensations as palliatives for inherently wrong lighting angles (top light, steep keys, side light on full-face shots). It is far better to correct the basic lighting treatment by repositioning the offending lamps.

Lighting men

In lighting men, slightly steeper lighting may provide more emphatically modelled, 'manly' results. Excessive backlight should be avoided, especially on bald on thinly grown heads. Lighting must not glamorise. On the other hand, long nose shadows, black eyes, hot ears, bright foreheads, glowing nose-tips, big neck shadows, etc., look no better on men than on women, and should be avoided.

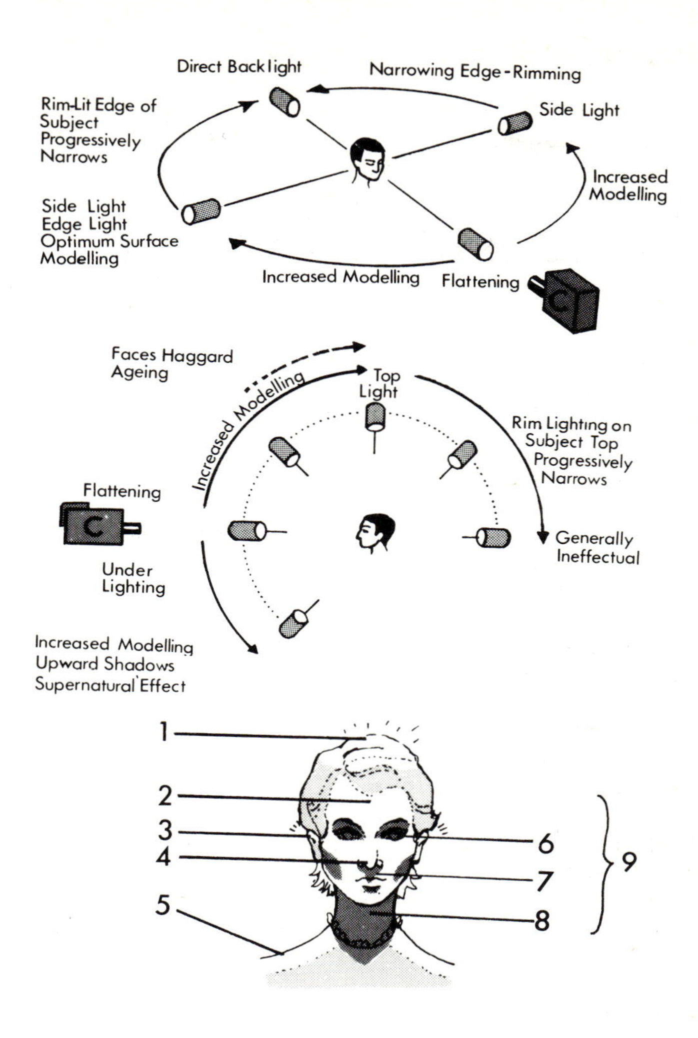

PORTRAITURE

Lighting angle
The effect of the lighting changes with its angle.

Pitfalls in portraiture
Ugly effects can occur through shadow formations, or overbright areas.
1. Bright top to a head. 2. Bright forehead. 3. Bright ears. 4. Bright nose.
5. Hot shoulders. 6. Black eyes. 7. Long nose shadow. 8. Black 'bib' from neck shadow. 9. Excessive modelling.
In this example, 1, 2, 3, 4 and 5 are due to steep bright backlight.
2, 6, 7, 8 and 9 result from a steep bright frontal key light.

Lighting a Single Person

Although we cannot lay down strict rules for portraiture, we can devise reliable working principles from which personal interpretations can grow.

All faces have fundamentally similar characteristics. We can, therefore, outline lighting treatment that will be generally appropriate for most people. Of course, our subjects' appearance may pose certain difficulties (e.g. deep-set eyes, protruding ears), and inconsiderate lighting could draw attention to, or exaggerate, these points. But the approaches outlined here do represent useful *modus operandi*.

The full face

The effect of a lamp is influenced by both its vertical angle (V), and how far round the subject it is located (H) relative to the camera's position. We need always to think three-dimensionally. The clock-face indicator helps us to pinpoint these lamp positions (page 36).

The *key light* can be located to the left or the right of the camera. Which we choose depends ideally on the inherent balance and proportions of our subject's features. In practice, it is often determined by production mechanics. A typical horizontal angle for the lamp to be positioned is between 15° to 40° round from lens axis, with 30° (7H or 7 o'clock) as a useful marker point. More offset (45°), and results are more dynamic (or more lop-sided). Less offset (0° to 5°) and the key is more youthening (or results flatter and less well modelled). The vertical angle of the key is typically from 5° to 45°, with a preferred range from 20° to 40°. The 30° direction at 2V (2 o'clock) is a guide.

Fill-light is effective offset between 5° and 30° (6H to 5H) with a height from level (3V) to +20° (+2V).

Backlight is usually slightly offset from a 12H position, round to 1H (and/or 11H). More widely angled, and defects arise such as white nose, black eye, and similarly crude modelling. These directions will be mirrored, of course, for a right-hand key.

The profile

Generally speaking, as the head is turned away from the lens axis, we move the key light, fill-light and backlight round correspondingly. If we think of keying 'along the nose' (−15°/+25°), we shall find this a useful guide. If the modelling is too strong, bring the key closer to the nose line. Where the result is too flat, take the key further round towards the back of the subject. If the nose shadow is too pronounced, move the key towards the nose-line or lower it.

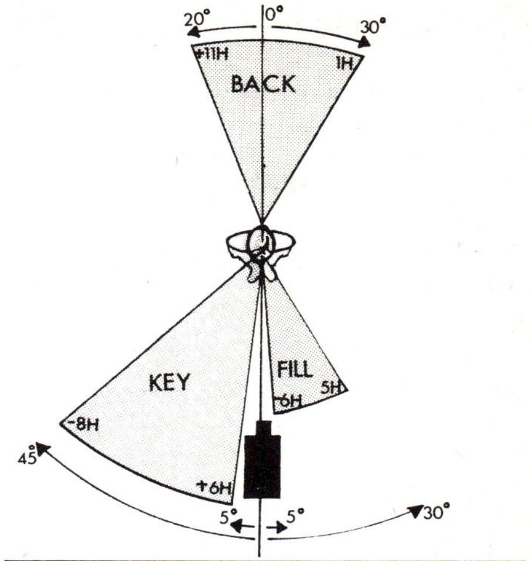

The full face
Here typical horizontal angles are shown, within which the key, fill and back lights are located to illuminate a *full face*.

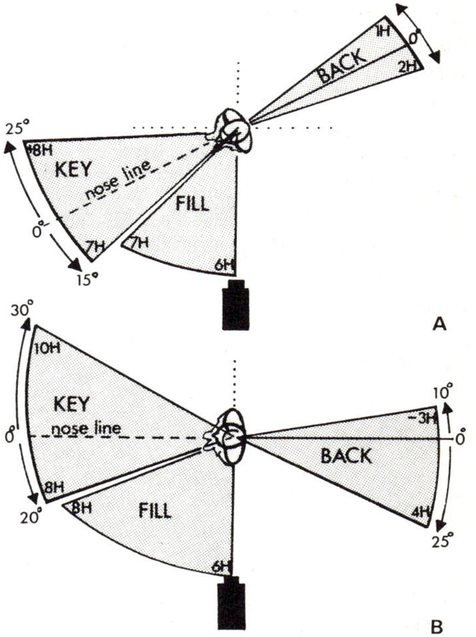

The offset head
For each fixed position that the head turns to (away from the camera), the direction of the key light alters correspondingly. Here we see typical lighting regions for a ¾-frontal face, A, and for a profile, B.

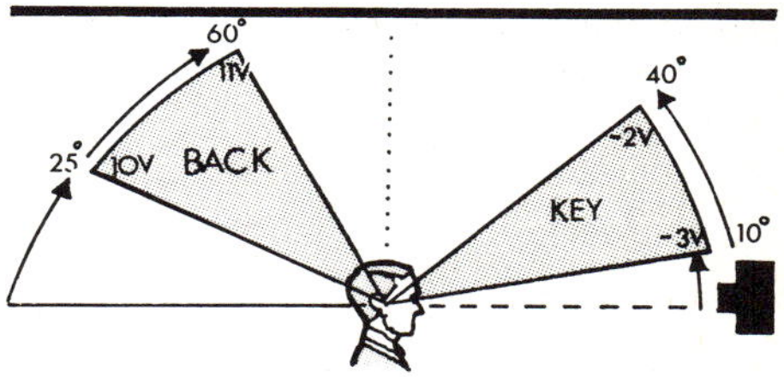

Vertical lighting angle
The steepness of both the frontal key light and the backlight should be controlled for optimum results.

Lighting Two People

As you might expect, there are various schools of thought here. Each of the techniques has its particular artistic or mechanical merits. Choose the approach that satisfies your own taste – and your circumstances.

Light direction

There are three favoured directions for lighting a pair of people. *Upstage cross lighting* locates keys and backlights behind both subjects. This is a regularly used studio system, that avoids shadows of the subjects or a sound boom on the background. Some exponents prefer *frontal cross lighting*, particularly where space is limited, as the key lights also illuminate the background; a handy expedient if one is short of lamps. The *aligned lighting* concept seeks to ensure that the light direction appears more consistent in intercut cross-shots. In each method, one can use communal or separate soft-light sources for the performers.

Divided backlighting

Ideally, we would light two people quite individually, each according to his own requirements. But in practice, conditions often preclude this. When two people are speaking together, they are usually about 0.6 to 1.4 m apart (2½ to 4½ ft). Especially at closer distances, it can be impossible to keep their respective key and back lights from spilling on to the other person. Any attempt to do so by steeper lighting simply degrades the portraiture.

Consequently, instead of the preferred *separate treatment*, we often have to accept that *communal treatment* is unavoidable. The closer the people are together, the fewer are the lighting opportunities. In very close proximities, it is inevitable that one person shadows the other. The quickest and usually the most rational remedy to clear the shadow is to reposition the people slightly (page 68).

In a communal arrangement, the key light for each person falls upon his neighbour as a backlight. Economical but inflexible, this combined illumination may suffice if the people have reasonably similar requirements. But fair hair, bald heads, light clothes, bare shoulders, all need less backlight. On the other hand black hair, curly hair, black clothing, velvets, require more backlight. Face tones, too, vary from some 30% to 40% reflectance and so require different key light intensities. At best, therefore, the success of communal light intensities can be a matter of luck.

However, one can often improve the situation by reducing the key or the back light area of the lamp beam with a piece of diffuser material (page 48). Where lamp-sharing is inevitable, this expedient can provide optimum results.

TWO PEOPLE

Camera positions
Typical *camera* positions used for a two-shot situation.

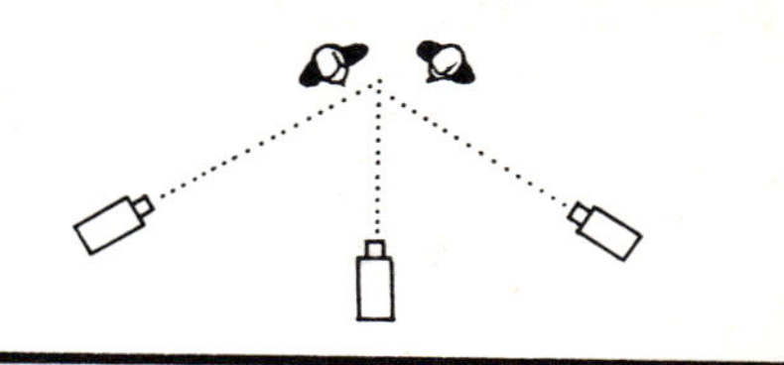

Upstage cross-lighting
Keys and backlights are located beyond a line joining the two people.

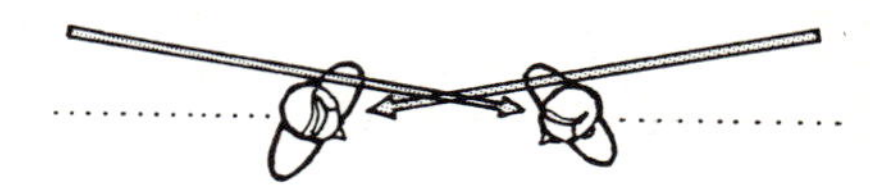

Aligned lighting
Key lights and backlights are aligned diagonally across the performers.

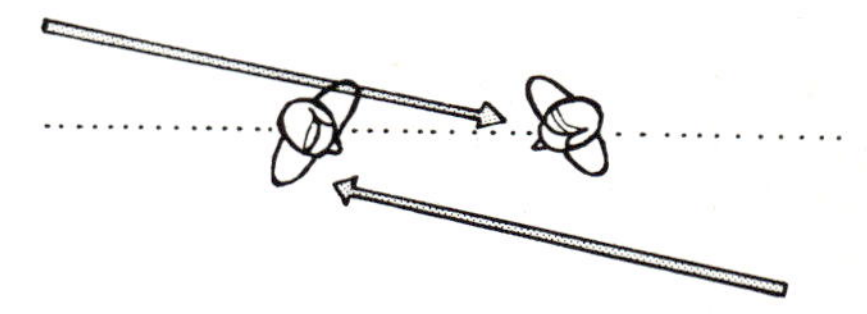

Frontal cross-lighting
Here the keys and backlights are located on the camera side of a line joining the two people.

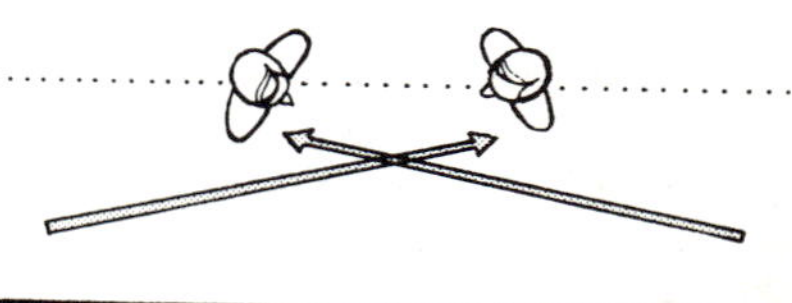

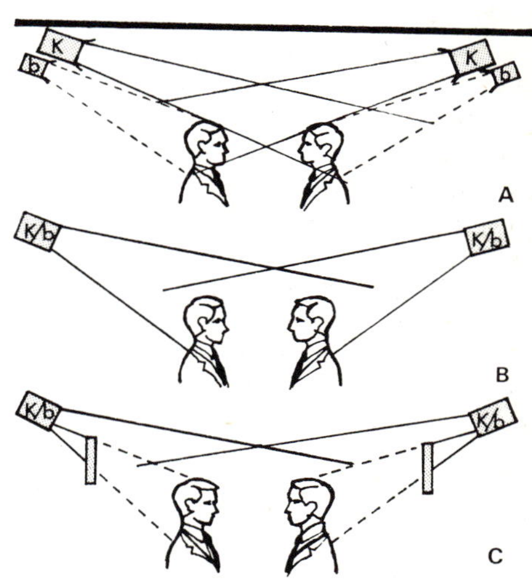

Divided backlighting
Methods of arranging backlight for two subjects:
A. Separate treatment.
B. Communal key/backlight.
C. Localised light reduction.

Lighting a Group

'Groups' come in all sizes, from a three-handed discussion to a large crowd. And yet we shall find that the underlying philosophies in our lighting approaches remain similar. At one extreme, we can light each person individually. With a small group of up to about half a dozen people, this is quite feasible. At the other extreme, it may be possible for us to light them *en masse*, with a communal 'three-point lighting' layout. Within these extremes, we have the concept of lighting small 'subdivisions'.

Individual treatment
Where people in a group are well-spaced, in a line (a panel-team in a quiz, or speakers behind a long desk), it may be practicable to light them *separately*. Each person has his own low-power backlight and a key light. Mutual fill-light will be necessary, as soft light cannot be confined.

Communal treatment
If our small group is closely-packed together, we may light it *as a whole*, with a single powerful key light, one or two backlights, and an assembly of soft light.

Subdivision
In this approach, we divide the assembly into a series of 'sub-groups' containing, perhaps, two or three people. Each group has its own shared key light and shared backlight, with fairly localised fill-light. It may not be possible to barndoor these lamps precisely to confine them to the allotted areas, but a limited amount of spill onto adjacent sections may not matter.

Such subdivision is a frequently used technique. It reduces the number of lamps needed, where space or equipment preclude more individual treatment. In an orchestra, we can light groups of strings, woodwind, etc., in subdivisions of four to six players. (Instrumentalists such as a pianist or harpist may still have to be lit individually.) A large audience can be divided similarly into manageable sections.

Although this lighting method precludes adjustment for individual group members' needs, it offers greater flexibility than the overall communal treatment. As with dual-purpose lighting (page 56), the conditions that are right for one person may not suit his neighbour. Video adjustment may improve matters (page 132) for *close shots only*, by adjusting exposure and black level to optimise each person's portraiture.

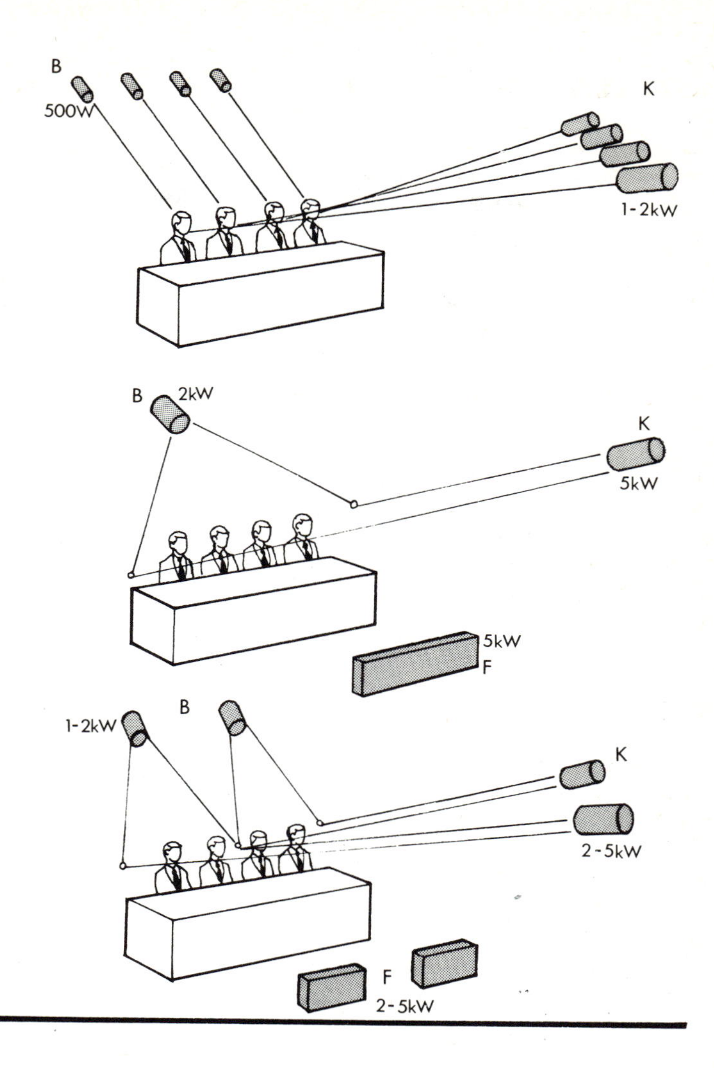

A GROUP

Individual treatment
In this regular group situation (panel for discussion or quiz), each person is lit individually. (The fill-light may be provided by localised or communal soft light units.) Typical lamp ratings are shown.

Communal treatment
Here, all the group shares the same overall three-point lighting. The directions and intensities may not suit all members.

Sub-division
Here a key, backlight and fill-light are shared by each pair within the group.

People Talking – Typical Set-ups

You cannot light with a tape measure! No description of a lighting set-up can hope to be more than a guide. After all, two people can switch on the same set of lamps and, by slight readjustments, produce substantially different results. Individual alterations of the lighting balance alone can change the entire pictorial effect.

Measuring lamp positions

Creative individuals are rightly suspicious of any 'statistics'. But such data have their places. It is far better to set out on an exploration with map and compass than with a wet finger in the wind!

The fundamental behaviour of light is predictable. Human heads are not so dissimilar that each needs radically different lighting. We can take a 'standard' lighting set-up, and sit anyone in the same position, and they will appear more or less satisfactorily lit. They may even look good. But, of course, such treatment does not take into account the idiosyncratic features of each individual, or show their personal characteristics to full advantage. It is a portrait, not a character study. Nevertheless, a lighting set-up of this kind can produce perfectly acceptable results, and can be modified as the occasion requires. Even in network studios, last-minute guests have to be catered for in exactly this fashion.

The diagrams opposite are essentially *guides*. They are measured indications of how you can start to light such subjects. Your particular circumstances will differ – but not in essentials.

Practical considerations

There are currently such variations in lamp design and efficiency, that no one type-quote would assist all readers. The luminants themselves cover a range including tungsten-halogen, and overrun sources, with widely differing outputs.

As a practical indication, typical operating heights and distances for lamps are quoted here. If you find their ratings inadequate or excessive for your cameras, alter their scale or the lens aperture accordingly.

A final word. The light output of lamps and light fittings are quoted by their manufacturers. But the passage of time (dust, tarnishing, ageing bulbs, etc.), can modify their performance. A replacement lamp may change a fitting's effectiveness considerably. Larger tungsten lamps contain granules with which we can scour off any internal carbon deposit from their glass envelopes, and so improve the light output. Supply voltages may vary. So we see that there are many variables at work to upset any hard-and-fast 'statistics'!

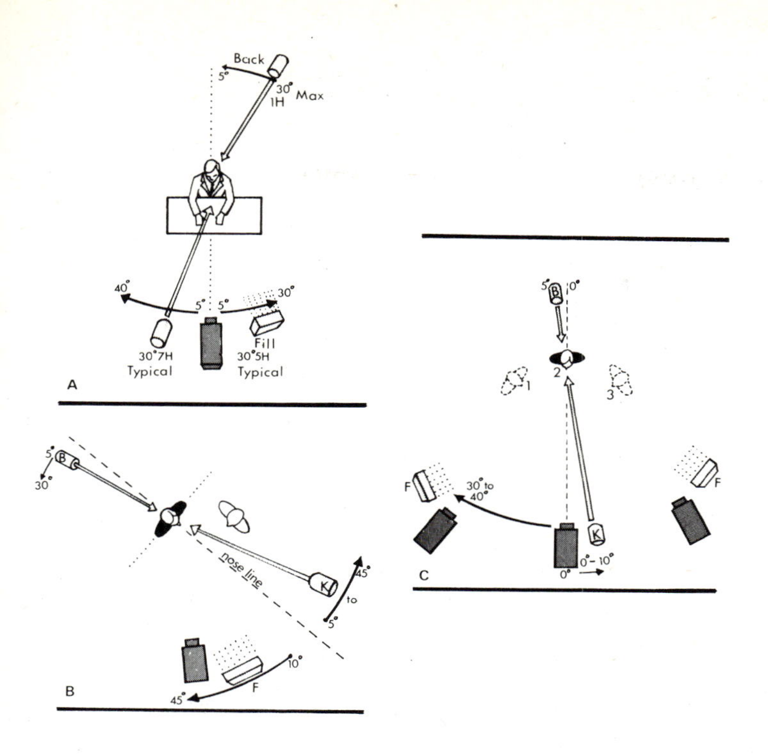

PEOPLE TALKING

A. Announcer set-up
Typical lamp positions, using lamp heights around 2.5 to 3 m/8 to 10 ft.

	Sitting	Standing
Key (2 kW)	2.5–3.6 m 8–12 ft	1.3–2 m 4–7 ft
Back (500 W)	1.5–2.5 m 5–8 ft	1–1.5 m 3–5 ft
Fill (1 to 2 kW) (at height 1.3 to 2 m/4 to 7 ft)	1.5–2.5 m 5–8 ft	1.5–2.5 m 5–8 ft

APPROXIMATE CURRENT CONSUMPTION OF LAMPS (AMPS)

	10 kW	5 kW	2 kW	1 kW	750 W	500 W	200 W
115 V	87	43½	17½	9	6½	4½	2
230 V	43½	22	8½	4½	3½	2½	1

B. Interview set-up for two people
Lamp distances as in A (height 2.5 to 3 m/8 to 10 ft).
Angles measured from the nose line (at right-angles to the body).
The other person is lit with an identical, mirror-image set-up.

C. Interview set-up for three people
Persons 1 and 3 lit as for a two-person interview (see B). Central person lit as in A, with a central key.

Problem Occasions

Even in the most straightforward productions, we can encounter situations that pose lighting problems. The solutions are by no means obvious, but with anticipation, satisfactory results can be achieved. Here are some regular examples.

The head turns

When a head turns through a right angle, the lighting that produced the most attractive results for the original head position may prove quite unsuitable for the other. For a profile shot, we know that the best key-light direction is close to the nose line, but if the speaker now turns to face the camera, his key (e.g. 3H or 9H) now provides unattractive side light. In such a situation we would normally bisect the turning angle, placing the key at a mid-point (45°) between the two head positions.

As a rule of thumb, we can say that where a head turns away from a key, portraiture generally deteriorates, whereas when the subject turns towards the key, the result is usually quite satisfactory.

In lighting a group of three people, we shall find this problem to some degree, with the central person, as he turns his head from a frontal position to face his companions on either side. Here we have the choice of angling the keys to suit, or providing different lighting for his various positions, or using auxiliary lamps to improve any shortcomings in portraiture.

Positions near walls

When people are positioned near walls, we shall either light them 'along the walls' as in the first example, 2A, or use a cross back key (¾-back key) as in the second, 2B. If the person turns between A and B situations, an auxiliary key may meet the case (see Figure 2B).

Where two people stand face to face near walls, lighting may be impaired. Avoid any temptations to squirt light down vertically from the wall top. It will only result in unattractive top light. Instead, we can adopt the method shown in B, so that the 'auxiliary' lamp now becomes the newcomer's key (see page 68).

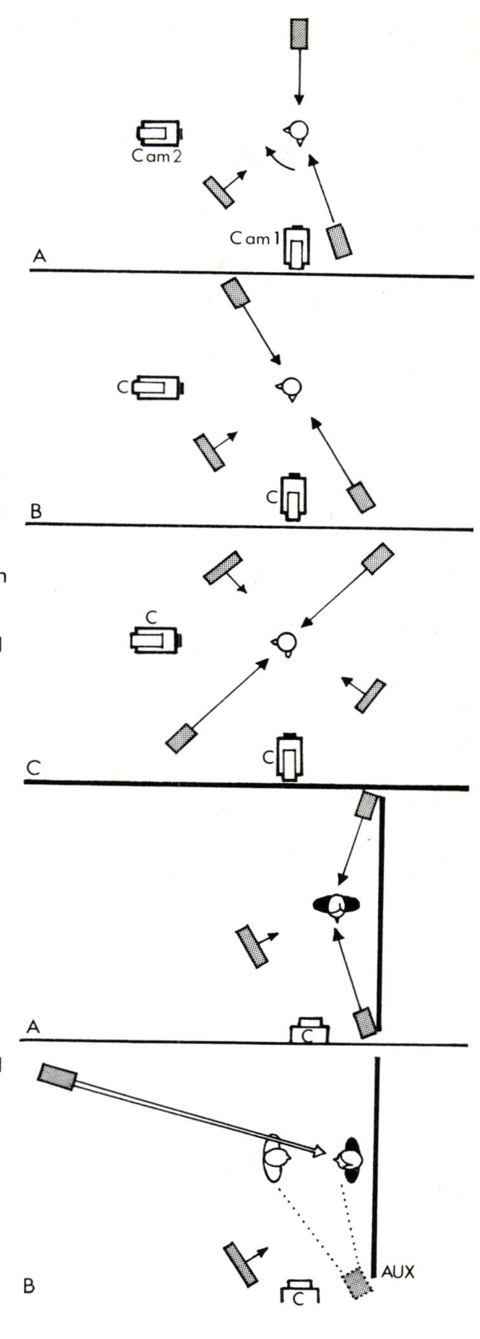

1. The head turns

A. The lighting set-up suitable for one head position may not suit another. On Cam 1 the full face is well lit, the profile flatly illuminated. On Cam 2 the head is unattractively bisected with light.
B. One lighting set-up may suit both cameras sufficiently for full face shots.
C. Another set-up suits full face and profiles on both cameras.

2. Positions near walls

A. Lighting along walls. Avoids illuminating the wall surface.
B. Upstage key light; supplemented by auxiliary key if subject turns to camera, or for two-shot.

Anticipating Trouble

Although it is in the interest of all members of a TV production team to plan and rehearse their show fully before it goes on the air, there are inevitably occasions when there is little opportunity for these normal procedures. Instead, the director and the rest of the crew have to anticipate. An outline plan of campaign provides an operation framework.

Lighting approaches
Fortunately, we can usually light the staging background for the production with reasonable certainty, and check this on camera beforehand. A dummy rehearsal with 'stand-ins' enacting the probable action, enables us to line-up shots. There are snags, of course. Their clothing may not be representative in tone or colour of that which the actual performers will wear.

Where action consists of people sitting talking in pre-arranged chairs, we can provide orthodox portrait lighting – remembering that eventually we may need considerably more or less light intensities than for the casual stand-in. A swarthy bearded stand-in's lighting requirements may not suit the delicately-modelled, bare-shouldered blonde who sits there during transmission.

Occasionally, too, one encounters the eccentric interviewee, who looks downwards nervously throughout, or the fidgety type who alternately slouches and stretches (getting dark eyes and a 'hot nose' respectively). It is not unknown for subjects to turn, looking away from the interviewer altogether, as they watch themselves on a nearby picture monitor in the studio! It is as well to be prepared for such eventualities, by rigging an alternative key (or fill-light) to accommodate the unexpected.

Lamp positions
Do not make lamp positions too critical, so that the portraiture is only good if the person sits still and at the exact angle. He may move in and out of his key light. He may tilt his head back, so that the back light strikes his forehead and nose. This happens particularly when we have made the backlight relatively steep, to prevent its lighting the interviewer who is sitting opposite.

Try to anticipate, too, that the director (or the cameraman) may, in the excitement of the transmission, take good shots that had not been foreseen. The most likely example occurs where people turn themselves into a face-to-face position, instead of the angled arragement that had been anticipated. Then, additional cross backlight (which now behaves as a frontal key light), may be needed to improve the new situation (page 63). If there is any likelihood of this particular dilemma, it might be worth providing them – just in case!

For larger area action, anticipatory lighting will normally follow the principles outlined in page 98.

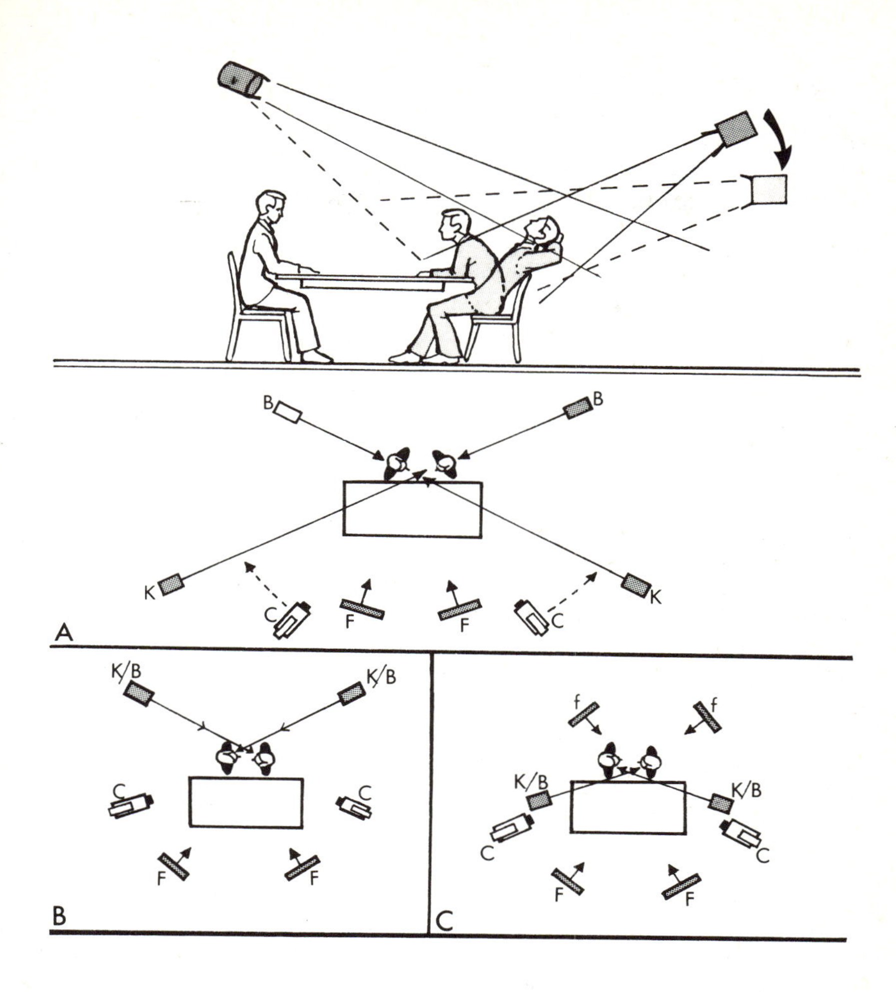

ANTICIPATING TROUBLE

Fidgety performers
Don't make lighting angles critical. Here in the forward position the subject has moved out of both key and backlights. Leaning back, the steep backlight emphasises facial modelling. The broken lines show improved versions.

Precautionary measures
Although the planned seating positions were *angled* (for optimum shots) as in A, the people may actually turn *face-to-face* instead! Cameras can re-locate themselves to correct shots, but the lighting no longer suits the new head angles. The upstage sides of faces are underlit. There are two solutions: B, cross-fade to different key directions, or C, fade up standby fill-lights F.

Lighting a Moving Person

Movement brings with it both disadvantages and advantages. The more people move around, the less opportunity will we have for exact light positioning. (If they remained still, we could light them precisely). However, while a performer is moving, there is less opportunity for the audience to become aware of any imperfections in our treatment!

The techniques we use to accommodate movement are allied to those we saw earlier (page 58) for lighting groups of people, namely: individual, subdivision, and large-area treatments.

Individual areas

Of course, we could try to follow the moving subject with mobile lighting (e.g. a camera light, a hand-held lamp) or a following spotlight. In certain circumstances such methods are quite appropriate, as an effect or as an expedient. But for the most part, these are not very convenient approaches.

Instead, we can light each *location point* with its own separate three-point set-up, *for that particular person's action*. As our speaker stands by the desk he is appropriately lit. Then, when he moves over to the blackboard, he enters a new set of suitably-angled lamps. Where we have different kinds of action in the same area, and different lighting is required for each, we may be able to cross-fade from one set of lighting conditions to another, either during a cut-away shot to another item (e.g. caption), or surreptitiously, during the action itself.

Subdivided areas

Following on from this idea of location points, we can treat movement by breaking the acting area into a series of subdivided lit regions. Now, *anyone* moving around by the door, the table, or the bench becomes lit by the lamps set for these areas. Although the same local treatment may have to suffice for quite different people in an area, action itself if often repetitive, and a rational lighting set-up will usually produce very satisfactory pictures nevertheless. It may be possible to adjust the lighting balance (by dimmers) to optimise for individuals.

Total treatment

The simplest ways of lighting for action are necessarily the least tractable. They involve lighting with an overall pattern that we *hope* will suit the action as seen from various camera viewpoints. We can modify or augment the treatment, as rehearsal indicates is necessary. For widespread or unpredictable movements, this is the only practical approach.

Individual areas
Each person's position is individually lit.

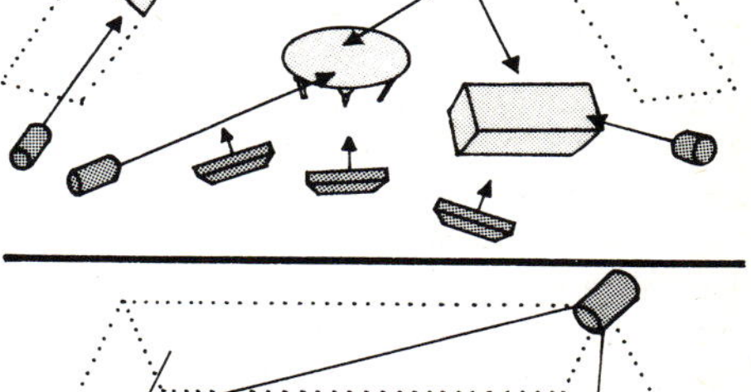

Subdivided areas
Each part of the setting is lit for any performer *in that area*.

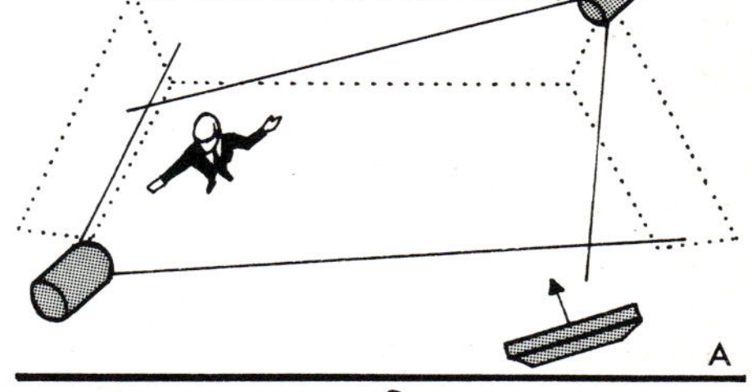

Total treatment
A. An *overall* three-point scheme to accommodate action.
B. *Dual-key* lighting splits the total area into two, each with its own three-point treatment.
C. *Soft frontal* area lighting, in which two ¾-backlights provide hard modelling for cross-shooting cameras.

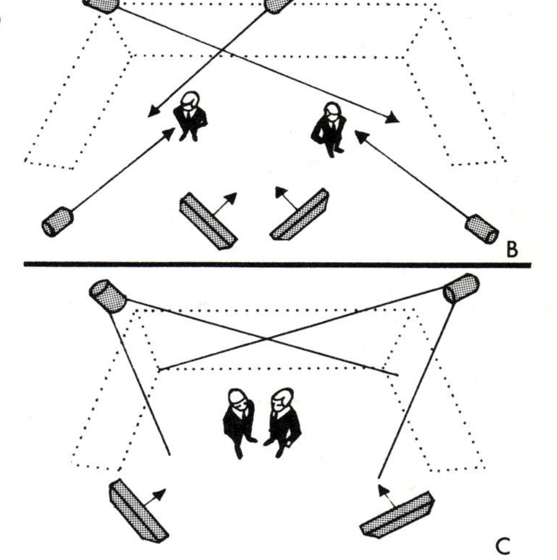

Positioning People

Where and how a person is positioned can have a considerable influence on our lighting techniques. Certain situations are an anathema to good lighting. Low ceilings, confined spaces, positions up against walls or shaded off by local structures all frustrate good lighting quality. It may be possible to introduce some sort of compensatory lighting (e.g. camera-light) that enables us to produce acceptable results despite the restrictions. Otherwise, the best recourse is for the director to change the performer's position.

Looking down

A downward tilt of the head has the same visual effect as steepening the angle of the key light. Facial modelling is emphasised, and 'black eyes' develop. We can anticipate this by keeping our key light lower, or by introducing a low fill-light. Otherwise, we may simply ask the performer to keep his head-up!

Close positions

Where people stand close together, we often find that one shadows the other. Sometimes drawing their attention to the shadowing will enable them to avoid it. Shadows may be 'cleared' if one of the people moves his head slightly, or the pair are angled to the light. The only lighting solution may be to move the key light – usually over a considerable horizontal angle – with the likelihood of upsetting the lighting of other nearby areas (page 56).

Against light surfaces

Whenever someone stands against a bright background, his face invariably reproduces much darker than usual. There are several reasons for this. The camera lens has to be stopped down to prevent the light-toned surface from being grossly over-exposed (page 132). If we try to lighten the now under-lit face by illuminating it more strongly, the chances are that this light will spill onto the background, making it even 'hotter'. An optical illusion known as *simultaneous contrast* (spatial induction) causes tonal contrasts to appear exaggerated. And, finally, the TV receiver itself may actually reproduce the face as a darker tone due to the electronic quirk of *black level variation*.

On balance, it is better to change the background tone, alter the camera angle to miss the surface, or move the subject to permit him to be lit separately from his surroundings.

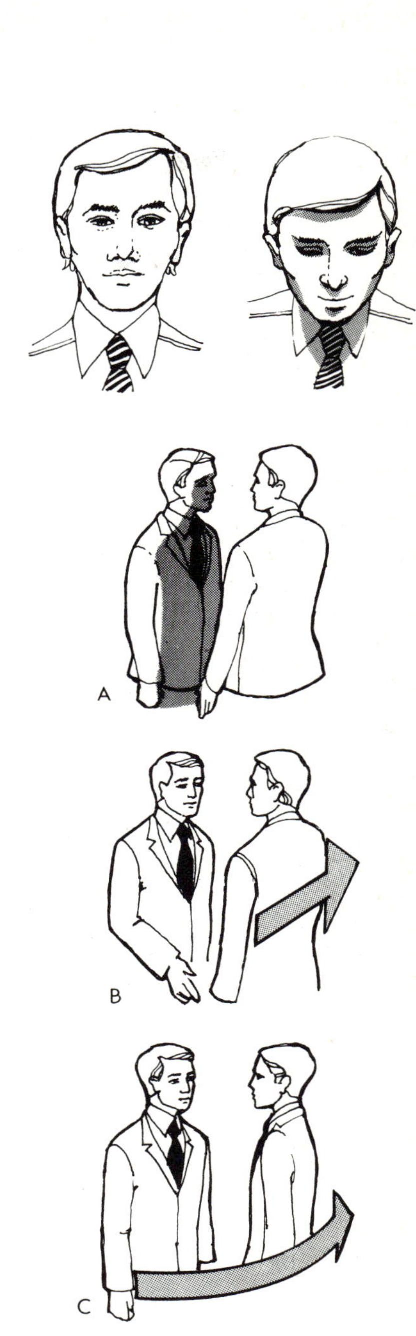

Tipping the head
As the head tips down, the key light effectively steepens, modelling becomes harsher, and downward shadows grow longer.

Clearing shadows
Slight readjustment of performers' positions will often overcome shadowing problems, A, either by sideways movement, B, or by pivoting their positions, C.

Changing Camera Positions

The visual effect produced by any given lamp will change with the camera's viewpoint. The diagram opposite shows you how striking this can be. From one camera position, a lamp can be dead frontal, and the resultant lighting quite flat. But move the camera round to shoot from another angle, and the effect of this same lamp becomes entirely different. The function of the lamp alters as we switch between cameras. This could mean that our subject could appear well-lit in one shot, and virtually unlit in another! How do we accommodate changes of this kind?

General solutions

We could try to light the scene from all directions – a ring of lamps – so that, irrespective of the camera's position, the subject would be illuminated. But the result would probably be a multiplicity of shadows, and very indifferent modelling as each lamp nullified the tonal gradation created by the next. The picture would tend to lack crisp, firm definition.

Far more satisfactory results can be achieved by using systematic lighting, following the principles we met earlier (pages 58, 62, 66). Area lighting, similar to that we used for a moving person, can be modified where necessary, to improve particular shots. This will usually entail adding an extra localised key light or filler for a specific camera angle.

Preferred position

In a *preferred position* approach, we ensure that the lighting for the more important position gives good portraiture, even if it looks rather less effective from another direction.

Dual set-up

Here lighting angles are carefully chosen to suit both camera directions well. It will accommodate equally successfully two people in conversation, or one person who turns through a wide angle to his new camera.

Altering the viewpoint
The effect of any given lamp changes with our viewpoint.

Area lighting
Here the area lighting set-up may serve both situations quite satisfactorily.

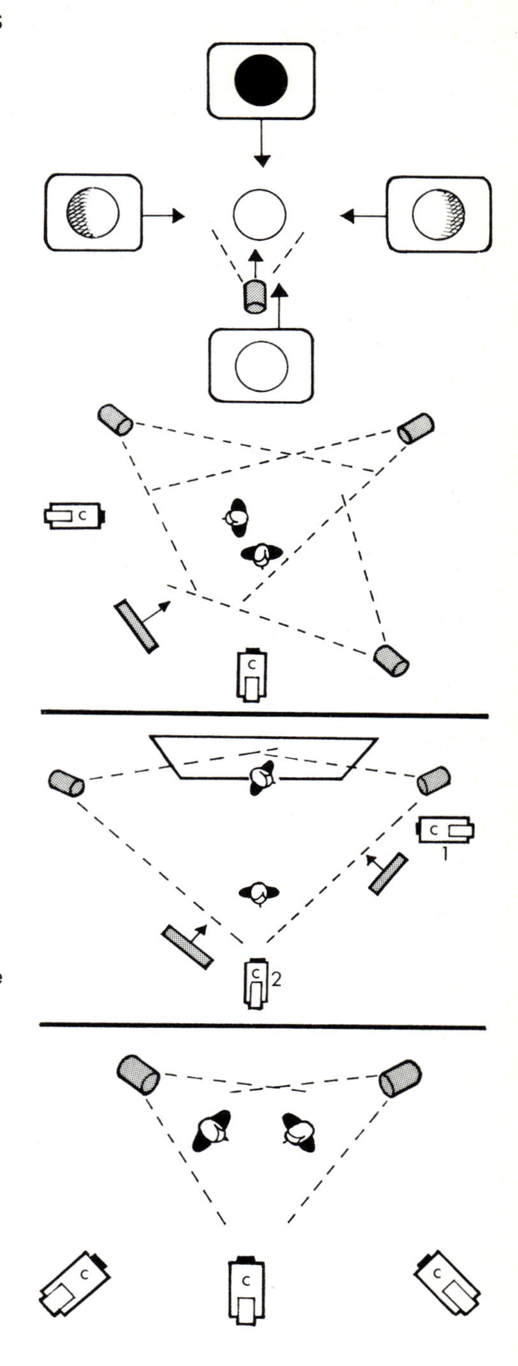

Preferred position
The set-up that produces best results for Cam 1 is still acceptable for Cam 2's shot.

Dual set-up
The localised lighting may suit both subjects equally, from all the camera positions.

Demonstrations – Lighting Problems

Demonstrations are a staple subject for the TV camera. They range broadly from formal lectures at the laboratory bench, to gastronomic revelations at the kitchen table.

Fundamental lighting precautions

The main drawback to a straight 'three-point lighting' treatment (page 36) is that the offset frontal key light may not suit both the demonstrator and his subjects. Its angle might not provide optimum modelling for items on the table. Subjects may shadow each other. The key light may also over-illuminate the front of the table.

Typical techniques

To improve the flexibility of this arrangement, we can barndoor the main key, restricting it to the demonstrator, and introduce additional localised lamps to the front and side of the table-top to light the display more effectively.

Where an overhead mirror is being used to provide a camera with top shots, we may also be able to utilise it to redirect a spotlight downwards, projecting appropriate modelling light into items on the table.

Backlight

Backlight is always liable to cast the demonstrator's shadow forward across the central working area of the table, causing obscuring or distracting shadows over the subject he is discussing.

We may be able to avoid this by using a lower-angle backlight that is barndoored to illuminate only the demonstrator's head and shoulders. Shadows from a low-intensity backlight may become sufficiently diluted by key lights to be quite unobtrusive.

Another approach is to use a soft backlight in order to avoid these shadows altogether, but it can give less effective lighting, and its spill-light is not so readily controlled.

Reflections

Sometimes backlight bouncing forwards towards the camera, or multiple reflections of studio lights, can prove troublesome. The only practical remedies, apart from removing or covering the offending surface, are to try repositioning it, to spray it with 'anti-flare' wax coating, or if unavoidable, to modify the lighting. Occasionally it may be possible to 'kill' a lamp for an isolated shot, to prevent a bad specular reflection. But otherwise we can only reduce the number of reflected light sources to a minimum (switching off lamps in other areas nearby) and keep soft-light sources in closely-rigged continuous arrays.

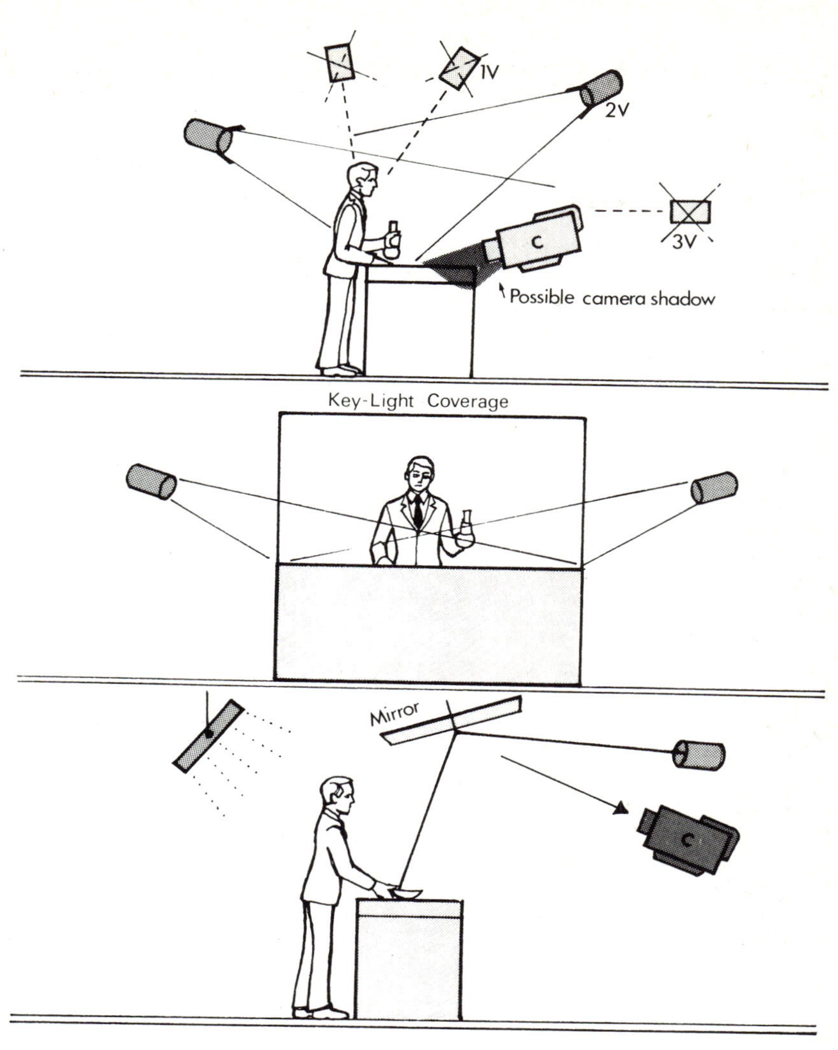

DEMONSTRATIONS - PROBLEMS

Lamp heights
Try to avoid a steep key (1 V) for demonstrators. (They tend to look downwards, and facial modelling increases.)
A shallow key 3 V throws shadows of close cameras over the table.
Backlights can throw forward shadows over the table.

Localised lighting
It may be preferable to confine the frontal key light to the demonstrator, and light his table separately.

Subterfuges
Overhead shots may be lit via the mirror the camera is using.
Soft backlight may avoid forward shadowing.

Demonstration – Typical Set-ups

A demonstration can be an exacting exercise, for it requires both good portraiture, and clear modelling of the subject being demonstrated (page 72).

The demonstration area

If we use a single key for both the demonstrator and the table, the results can be something of a compromise.

Particularly where the forms of the subjects being demonstrated are at all complex (e.g. machinery or equipment), it may be preferable to light them separately. In the extreme, we might need to reposition the demonstrator to facilitate this, having him stand beside his subject rather than behind it.

Sometimes we shall want to emphasise part of a display with a strong spotlight. However, we must not overlook the possibility that if the demonstrator puts his hand in this area, it will be considerably over-lit. Occasionally, it becomes necessary to isolate a demonstrated item entirely, in order to light it for optimum results.

Positioning the demonstrator's key light

Keep the key light down to a reasonably shallow vertical angle if possible (say 30°, i.e. 2V). If you make it flatter (between 3V and 2V), you may dazzle the demonstrator, or prevent his reading any cue sheets or reminder boards near the camera. There is a greater possibility, too, of *kick-back (hot spots, blooming)* from light-toned or glossy surfaces, and for specular reflections. We are liable to find also that the camera's shadow is cast over the table (page 72), particularly in closer shots, or where the camera elevates to look down to detailed action.

A steeper key light worsens portraiture, especially when the demonstrator bends over or looks down. In addition, it may introduce shadows from any overhead obstructions (e.g. a mirror).

The wall display

Wall displays in the form of maps, charts, or diagrams, are a familiar adjunct to many demonstrations. Care is needed here to prevent the demonstrator's shadow falling over the display (check if necessary, whether he is left- or right-handed) or to avoid multiple shadows and any spurious reflections from the surface of the graphics.

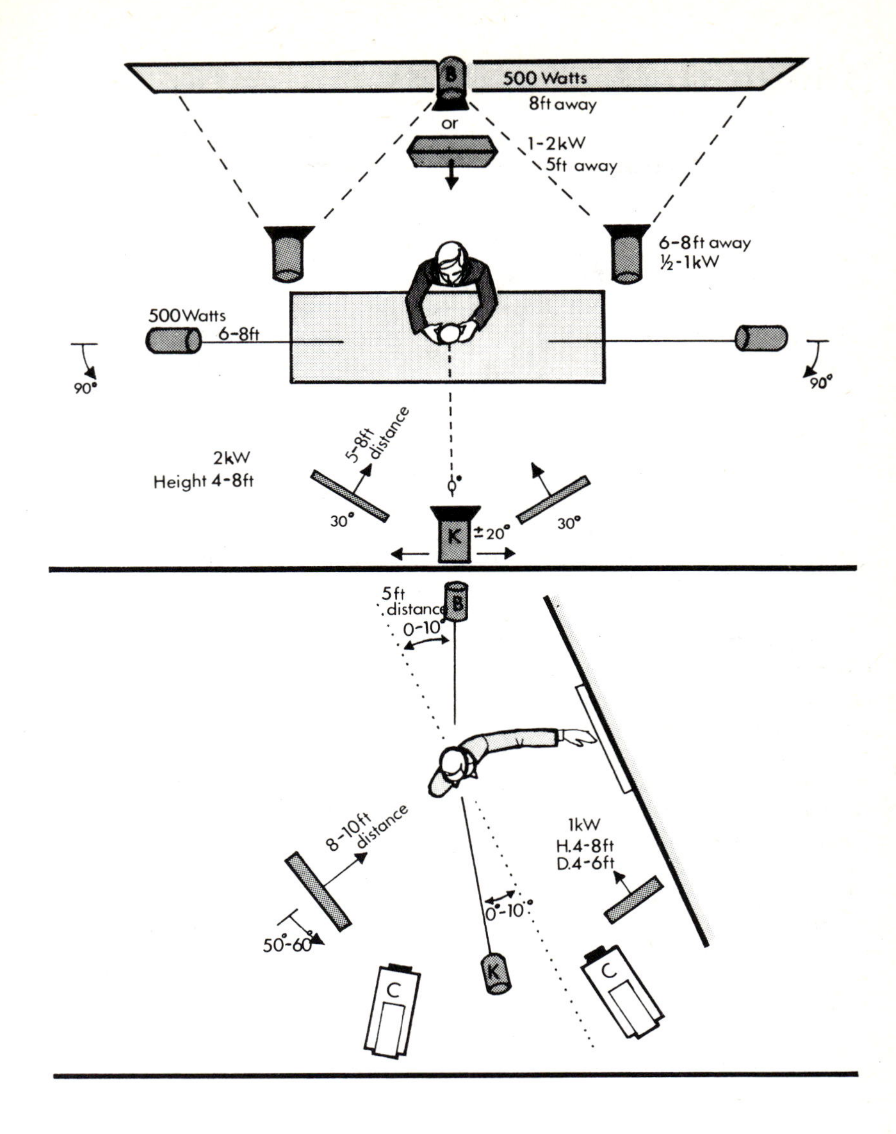

DEMONSTRATIONS – TYPICAL SET-UPS

The demonstration table
The key light here is relatively central, and the fill-light well offset to suit cross-shooting cameras.
Backlight may be hard (barndoored from the table) or a soft-light source.

The wall display
At a wall display, ensure that the demonstrator's shadow does not obscure.
Avoid backlight throwing shadows forward over the wall.

Lighting for Clarity

For many types of production, *clarity* is the essence; especially in the monochrome picture. But clarity is by no means just a matter of overall sharpness and lots of light. Planes can merge. Outlines can be unclear. Details may become confused together.

Sharpness

The depth of field in our picture is influenced by the lens aperture, and hence by the general light levels. With higher prevailing light intensities, the lens can be *stopped down* to a small aperture (e.g. *f*16) so that objects from close to distant planes are equally sharp. Where light levels are much lower, the lens has to be *opened up* to a large stop to provide correct exposure (e.g. *f*2), and the depth of field becomes restricted. Focused depth diminishes, too, the closer our subject is to the camera.

With a shallow depth of field, focused subjects stand out clearly against an indistinct background. No distractions arise from extraneous items nearer and further from the camera. But if this depth is insufficient for our purpose, any focusing restriction becomes frustrating, because we cannot see clearly enough.

A picture's effective sharpness can be reduced too, by flat or multi-shadow lighting blurring information (page 70).

Tonal differentiation

The tonal contrast between subject and background is important if the two are not to merge. Mostly, we want our subject to appear lighter than its background, for dark-toned areas tend to lose modelling and detail. Over-light backgrounds often distract (page 68).

Backlight can improve the outline clarity of objects, particularly where they are ill-defined (e.g. feathers, foliage), or are made from translucent materials.

Do not be deceived by colour differences between subject and background. In a monochrome picture, remember, quite different hues are reproduced as identical grey-scale values if they are the same tone.

Physical characteristics

Lighting should convey appropriate texture and form. An attractively-contoured surface could, under flat lighting, reproduce as an unmodelled, detailless plane. Conversely, a studio photographic blow-up that effectively simulates an exterior scene on camera when flatly lit, will be utterly unconvincing when any creases, undulations or spurious shadows destroy the illusion.

Shadows and highlights can camouflage or obscure information. This may be done deliberately for dramatic applications, but arising accidentally they can ruin clarity just where we need it.

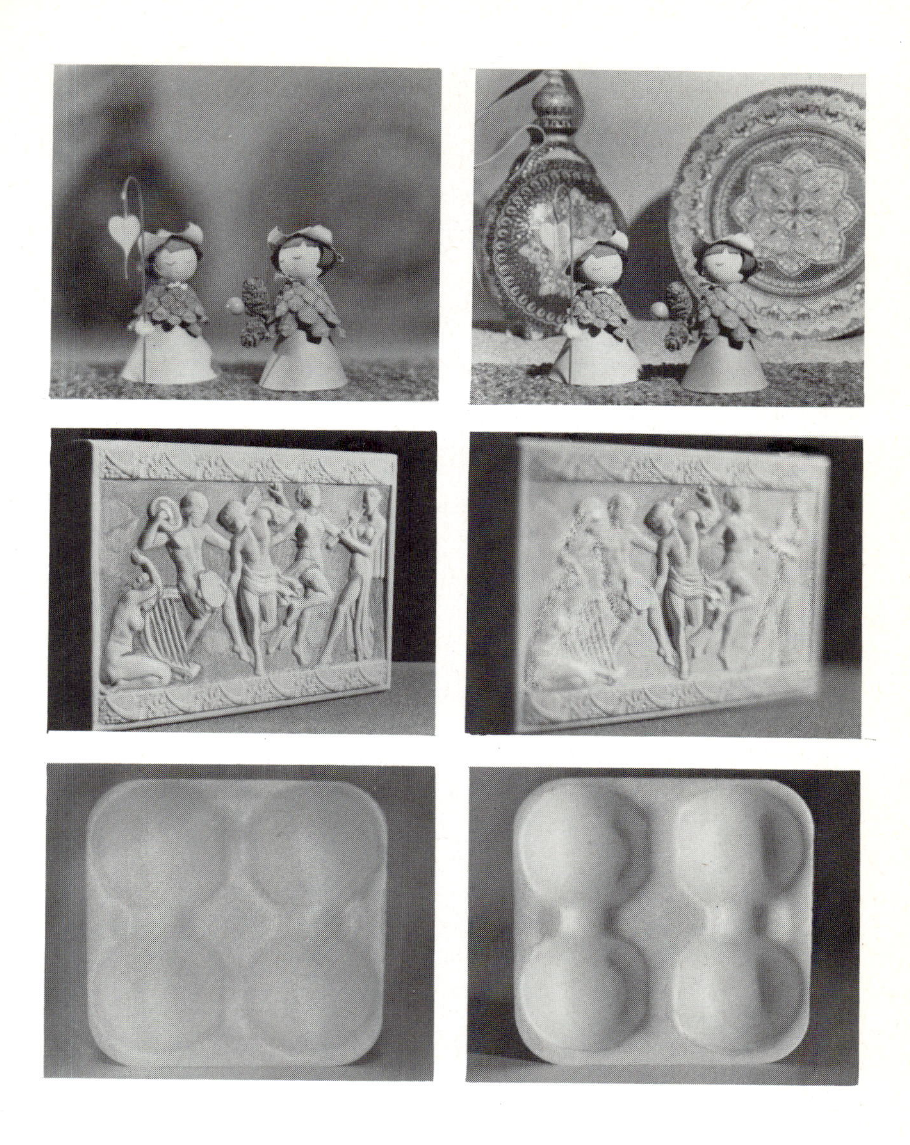

LIGHTING FOR CLARITY

Depth of field
By *restricting* depth of field, we can eliminate unwanted details, and concentrate attention. By using *maximum* depth of field, we can provide overall clarity.

Form and texture
Lit frontally, modelling is supressed. An angled key light reveals the true contours.

Distance, Space and Size

The interpretation of the flat TV picture is a remarkably subjective business. Irrespective of the visual clues present, our judgment of what we perceive there is greatly influenced by the tones we see, their area, and the tonal distribution. Tonal values can modify our interpretation of distance, space and size considerably. So here, clearly, the influence of lighting is intrinsic.

Distance and space

Our visual attention tends to move towards lighter areas in a picture. Where tones are graded around a light region, we invariably interpret the light plane as being further away from the camera – even when there are no perspective clues present in the picture. If there are, in fact, also perspective indications, the illusion of depth usually becomes very pronounced.

So, if we want to suggest that certain planes are more distant than others, we light them more intensely. Consequently, we usually illuminate surfaces that are intended to simulate distance – e.g. a sky cloth or photo-backing outside a window – to a proportionally higher level to enhance the illusion.

By lightening the walls of a room we convey spaciousness. Reduce the amount of light falling upon them, and the darker wall tones encourage an impression of closeness – or even claustrophobia. So we see how lighting can modify pictorial space; emphasising its extent, or bringing it crowding in.

Size

However irrationally, tone also modifies one's interpretation of *size*. As we increase the tonal difference between two otherwise identical objects, we find that the lighter will tend to look larger, while the darker seems smaller.

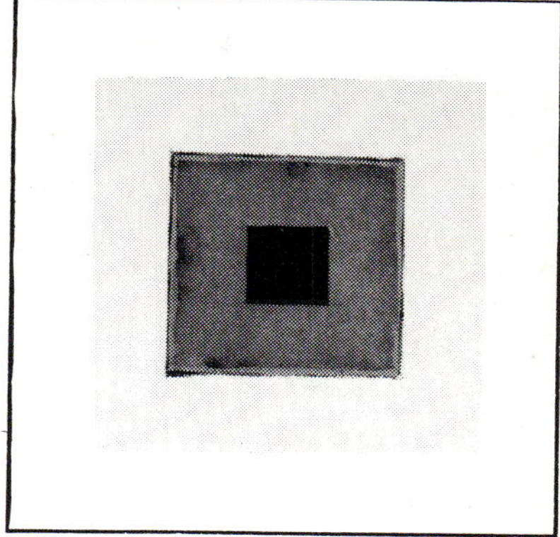

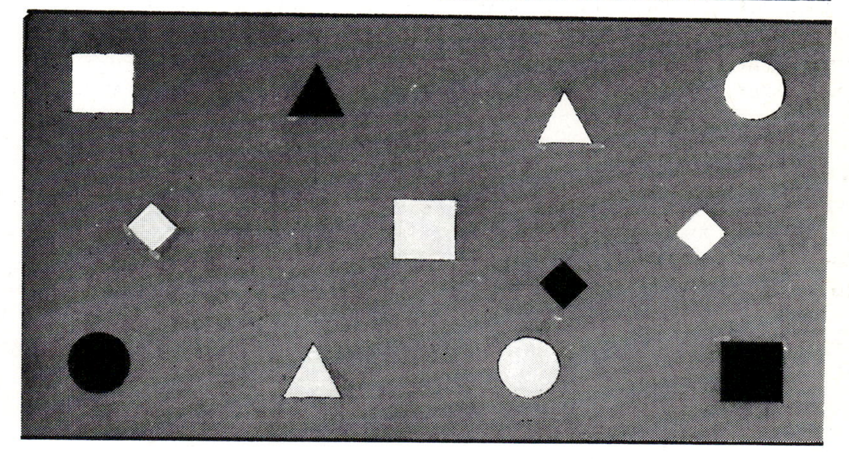

DISTANCE, SPACE AND SIZE

The impression of space and distance
Progressively lightened tones create the illusion of distance. Here we see that the figure on the left conveys a strong illusion of perspective and distance, entirely lacking in the right-hand design.

The impression of size
Lighter-toned subjects look larger than identically-sized ones of darker tone, irrespective of their shape.

Pictorial Lighting

Of course there can be no rules for creating pictorial appeal in our pictures! Nevertheless, experience does suggest various features that lead to attractive results, while others, conversely, provide less satisfying effects. Here then are some general pointers to watch out for when lighting the setting.

Walls

You will usually find that walls look more attractive if gradually shaded, leaving them darker at the top. For most purposes, we should avoid any sharp light cut-off on walls. Flatly-lit walls, or backgrounds that are noticeably brighter at the top, are less pleasing to the eye. Where face tones are light, wall tones are mostly kept lower, to achieve optimum clarity (page 76). Do not exaggerate wall texturing or modelling (e.g. with edge or top lighting), except for special effects. Try to avoid any hard structural shadows that destroy one's impression of shape and construction of the staging.

Shadows

By leaving foreground areas in comparative shadow, we can both attract attention to the more brightly lit distance, and enhance the illusion of depth.

A single, cast shadow can be extremely attractive. But we should generally avoid lop-sided shadow-formations, incomplete shadows, or disrupted patterns (e.g. where scenic structure interrupts).

Shadowy areas in a setting can be intriguing, mysterious. Shadows can add interest – but they should appear rational and purposeful. It is easy to over-dramatise or to introduce inapt shadowing. Then shadows merely frustrate, for they obscure detail the audience needs or wants to see.

Distorted or straggling shadows (e.g. from wall-light brackets) tend to be obtrusive.

Light patterns

Bright background areas that contain no intrinsic centre of interest can become unattractive and mislead the attention. Odd scattered, unco-ordinated light patterns on a background lack visual appeal, particularly if they cluster around the top of the picture. Geometric patterns must be firm and accurately delineated, or they draw attention to their defects. Defocused, abstract, soft-edged patterns can result in some delightful pictorial effects. Any decorative light movement is usually most pleasing when it involves slow, indistinct shapes. Fast, regularly-moving, sharply-defined light patterns soon appear mechanical.

PICTORIAL LIGHTING

Background shading
Gradually shaded backgrounds generally prove more attractive than evenly-lit or segmented background tones.

Complete shadows
If you are making a feature of a shadow, ensure that it is not broken up or cut short by staging.

Providing Shadow Effects

Shadows can be used in many ways; to stimulate the imagination, to conceal information; or, equally well, to reveal it. They can engender an atmosphere simply, effectively, and with economy. Shadow effects do not necessarily involve special equipment. Many can be derived from the simplest hard-light sources.

Types of shadow

Fundamentally, there are three main ways in which we can produce decorative shadow effects: cast shadows, projected shadows, silhouetted shadows.

* *Cast shadows* are the most obvious. We place an opaque object in the beam of a hard-light source.

* *Projected shadows* are derived from the focused image of a stencil or slide in a projection spotlight (profile spot) (page 22).

* *Silhouetted shadow* can be achieved either by positioning the unlit subject against a light-toned background, or by casting its shadow onto a screen from the rear.

Effective shadows

We need to be able to control the shadows we introduce, to give them the exact size, shape, and sharpness we are seeking. The following pointers will help you to do this.

* *Stability* is important. Unless the subject is to have movement (e.g. swaying leaf shadows) hold it rigid.

* *Dilution* from spill light can grey-out shadow formations and reduce their crispness.

* *Sharpness* results from a series of interrelated factors. We shall achieve the sharpest shadow from a small point source, using an object distant from the light, and relatively close to the background; the image being undiluted by stray light. The object and background should be at right angles to the light beam, and preferably flat.

* *Size* of the shadow is dependant upon that of the shadowing object, and its relative distance from the lamp and from the background. The nearer the subject is to the lamp and the further from the background, the larger will be the shadow. But notice that these conditions conflict with those for optimum sharpness. Clearly, we may have to strike an appropriate balance between size and sharpness. For projected shadows, the equipment's distance, its lens angle and slide area are decisive size factors.

* *Distortion* results whenever the shadowing object and the plane upon which its shadow is produced, are not at right angles to the light beam. Consequently, with the exception of central rear or frontal projection (page 114), we usually encounter some degree of distortion. This can sometimes be compensated, but quite often such distortion can appear natural, or decoratively effective.

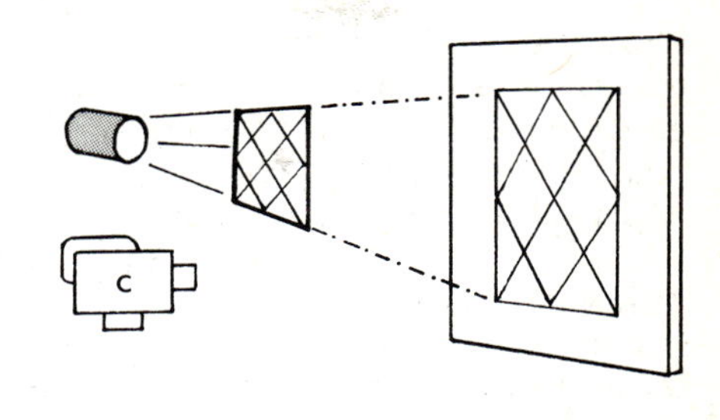

Cast shadows
Any opaque subject can be used to create shadows; including tree-branches, tracery, cut-out shapes, window frames, cookies.

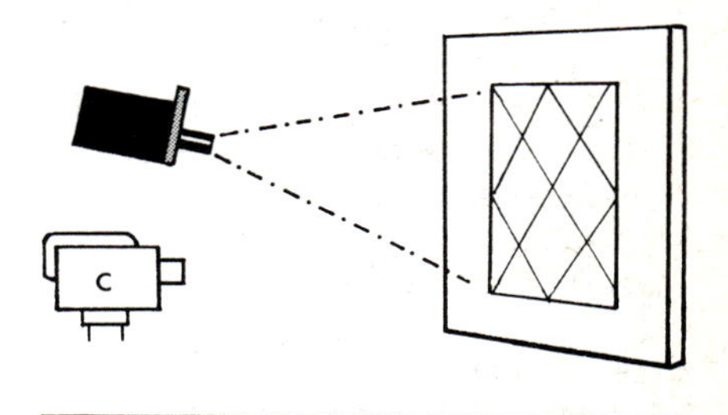

Projected shadows
Stencil masks can provide precise and elaborate patterns derived from a slide or a metal mask.

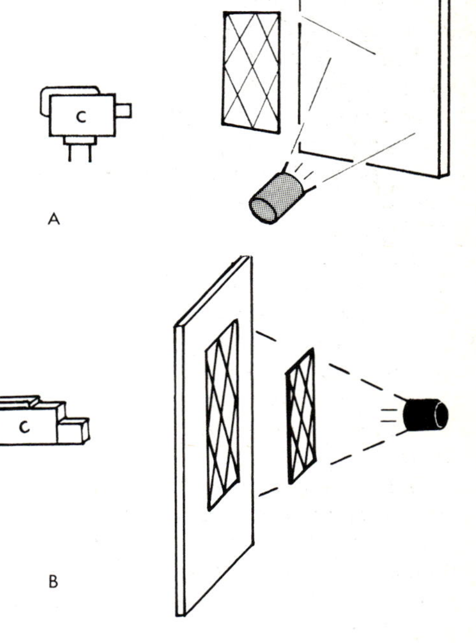

Silhouetted shadows
Emphasis is on outline shape, A, the unlit subject being shot in front of a bright background, or B, its shadow rear-projected onto the translucent background.

Controlled Shadows

Sometimes, although we seek firm, clear-cut modelling in our subject, and therefore use a frontal spotlight as a key, we dislike the associated shadow falling over the background. It disrupts or unbalances the picture, or distracts the attention. However, we do not want to resort to shadowless lighting treatment, for this would reduce texture and contouring in our subject. What then is the solution?

Moving the key light

The closer our subject is to its background, the more difficult it becomes to avoid casting a shadow upon that surface. So the initial step, and usually the best remedy where the key is angled, is to move the subject forward towards the camera.

If the key light is dead frontal, the camera will be unable to see the shadow, but lighting will be flat. As we progressively offset the key, the shadow moves out from behind the subject. Moving the key sideways will displace the shadow – in the opposite direction – and may broaden it. Raising the lamp height causes the shadow to be set lower on the background – but remember, subject modelling coarsens as the angle steepens.

So we see that shifting the position of the shadow, to hide it perhaps in some less obvious part of the background, or to throw it out of shot entirely, has its drawbacks. For the sake of removing the shadow, we may degrade the subject's entire lighting. In fact, if the subject is actually touching the background, its shadow will just spread beside it, and re-angling the key light cannot improve matters.

Restricting the light coverage seldom clears *subject* shadows on backgrounds. Because the unwanted shadow results from the subject itself, it follows that any attempt to shutter off the offending area, must cause that part of our subject to be left unlit.

Relighting the background

It is always a temptation to try to eliminate shadows by 'lighting them out' – i.e. diluting them with additional light. But this technique has its limitations. By the time the shadow has been sufficiently illuminated to make it unobtrusive, we have usually over-lit the background.

Disguising the shadow

Occasionally we can follow the opposite ploy, and hide our subject's shadow within a dark or broken-up area of the background. (Mic shadows can remain undetected amongst foliage shadow patterns.) Rear-illuminated translucent screens dilute shadows falling upon them – provided that their light tone is suitable for the shot.

CONTROLLING SHADOWS

Subject-to-background distance
As the subject is moved further from the background, its shadow moves downwards and becomes less obtrusive.

Moving the key light – raising
As the key light is raised, the subject's shadow on the background moves downwards – and subject-modelling increases.

Moving the key light – sideways
As the key light is moved sideways to the camera position, the subject's shadow is offset in the opposite direction.

Shadowless Lighting

Diffuse lighting has considerable charm – when appropriately applied. Spread around indiscriminately, soft light floods the scene and kills modelling. It reduces our impressions of solidity, can modify our interpretation of size, and of distance. But used effectively, shadowless lighting engenders delicate, etherial effects. Paradoxically, the subtle half-tones that soft light can provide may prove too subtle for the TV system to reproduce well!

Positioning the soft light
The secret of using soft light successfully as the main luminant, lies in our keeping it uni-directional, and angled to the camera. Diffused light from the camera position flattens the shot. But use this light source offset (say, 4H or 8H and 3V to 2V), and modelling forms. The shaded, unlit areas will often be sufficiently filled by reflected light or random light scatter, to require no fill-light as such. However, where we wish to avoid pronounced shading altogether, we may introduce a slight, carefully-regulated amount of filler also.

Applications for shadowless lighting
Completely shadowless lighting is effective for high key studies, for highly reflective subjects (e.g. silver, glass), and to create an atmosphere of airy detachment in large-area display. The general light levels should normally be kept high, for low-intensity soft illumination is liable to produce a dull, 'overcast day' effect, that offers little in the TV picture.

For many applications we shall not use truly shadowless lighting alone, but introduce hard backlight to show the outlines of our subjects. When carefully controlled, most beautiful effects can be achieved by this means. Classical ballet, for instance, may use directional soft light for frontal illumination, and a single central back light covering the acting area. (Upward-pointing lamps on the ground, light legs shadowed by skirts.)

Background lighting
Large banks of soft light produce a considerable quantity of heat. Moreover, it can become difficult to control their coverage. If we are aiming to illuminate a large, plain background evenly with soft light, we are quite likely to discover that the spill from adjacent sources builds up into bright patches that are not readily eliminated. When Fresnel spotlights or some other similar hard sources are employed instead to illuminate the background, we can close barndoor flaps or use flags to restrict overlapping light beams from such doubling (page 102).

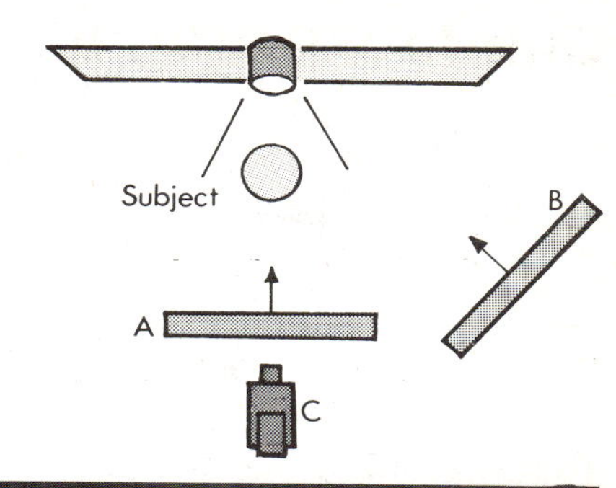

Soft-light position
When soft light, A, is located near the camera position, C, the resulting illumination is flat and the subject modelling minimised. But when offset as at B, delicate half-tone modelling results.

Hard backlight – soft frontal
The delicacy of offset soft light modelling can be enhanced further by the attractive rimming and shadow effects produced by backlight.

Scenic Lighting – Approach

When effective staging is skilfully lit, the results look *obviously* right, with strong visual appeal and the creation of a convincing atmosphere. But the mechanics which together build up that impression are themselves self-effacing. Such results come from methodical, developmental treatment that makes the best use of available time and facilities.

General methods

It is seldom sufficient to rely on the 'general illumination' from our filler or base light to illuminate a setting. The results would be too random, dull and unpredictable.

Ideally, each lamp we use should have its own intrinsic purpose. A key is arranged to light a performer, or to illuminate part of a wall, or to model a piece of furniture, as the case may be. In a smaller setting, or where facilities allow, this approach offers the optimum opportunities.

However, when such individual localised treatment is not possible, we may need to use dual-function lamps that serve both to light the performers, and to illuminate the setting as we see in the top diagram (also pages 56, 60, 62, 66, 70).

Specific background lighting

The best practice is undoubtedly to treat background lighting as a series of separate sections. This gives us close control over background brightness and modelling at any point, relative to its foreground subjects. Moreover, we can angle and shape the coverage of each lamp to suit the staging, and so develop the particular atmospheric environment the production needs. The more generalised the background lighting, the less dynamic is the pictorial impact likely to be.

A Fresnel spotlight is designed to give a soft-edged beam. Its flat even illumination is given a progressive intensity fall-off round its periphery. This enables us to blend beams edge-to-edge, to produce continuous, even illumination where necessary. It provides us too, with the opportunity to light the background in spots that fade off at their edges.

Using barndoors on Fresnel spotlights, we can isolate regions, and avoid light spilling onto nearby surfaces. The edge cut-off is softened off by diffraction. If we require a sharper border, we will have to use a flag or similar obstruction in the light beam.

The beams of hard-edged spotlights (effects spots) cannot be merged; although they may perhaps be aligned side-by-side to extend their coverage. Hard-edged light beams normally appear obtrusive, except in applications where they are environmentally appropriate as an effect.

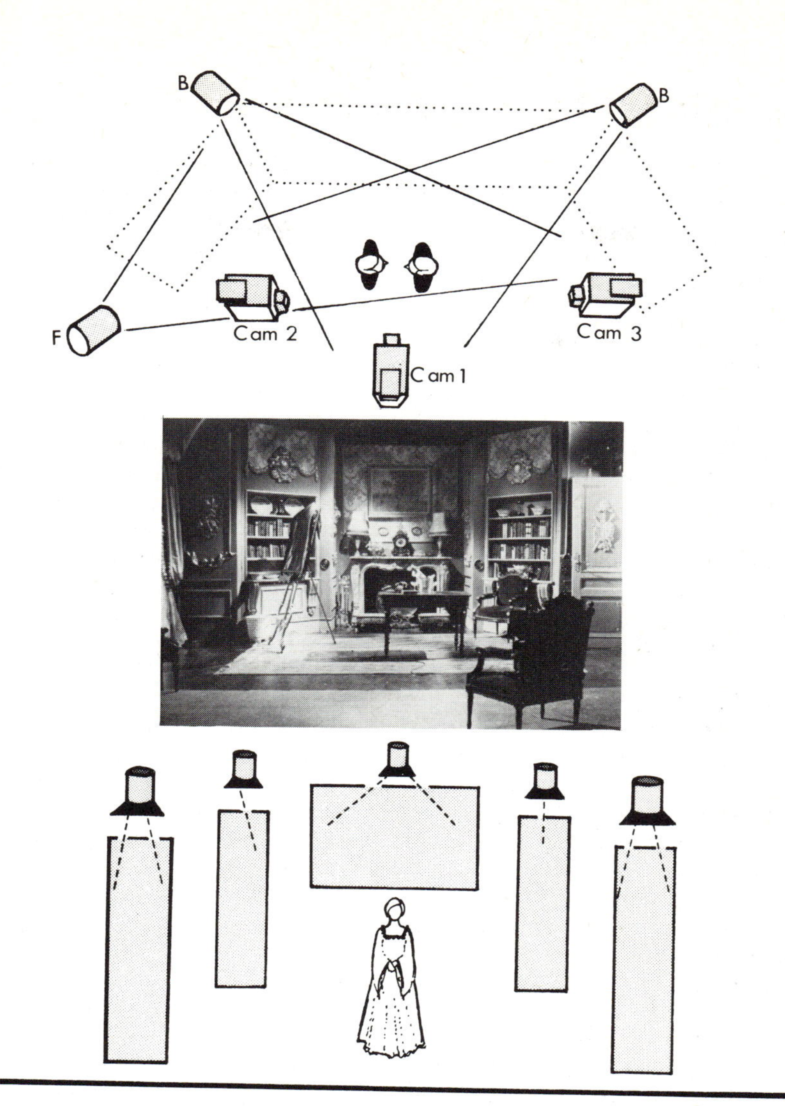

APPROACHES TO SCENIC LIGHTING

Dual-function lighting
The function of a lamp can depend on the camera angle used. The two lamps B serve as ¾-backlight for Cam 1, and as frontal keys for cross-shooting Cams 2 and 3. They may also illuminate the side walls. Lamp F is a frontal key for Cam 1, backlight for Cam 3, and also lights the rear wall of the setting.

Specific background lighting
The background can be lit in a series of conjoined areas, each merging into its neighbour. Alternatively, each section of the staging may be lit as a whole by a separate spotlight.

Scenic Lighting – Neutral Settings

Neutral settings are the backbone of economic staging. Their all-purpose, non-associative applications lend themselves to many types of production.

Plain backgrounds

Plain, unbroken backgrounds can be formed from runs of conjoined flats, from hung cloths, and from cycloramas. Lighting enables us to alter their appearance in a variety of interesting ways.

Typical background tones range from off-white, through shades of grey, down to black – although with light we can modify these values considerably. Light grey, for example, can be taken up to white and down to black under proportional illumination.

Plain pale blue or dark blue material can serve also as 'sky cloths' (for day and night respectively) for colour cameras – to save the need to use blue-coloured light, which might spill onto adjacent areas.

The plain background can be evenly lit, can be shaded (page 80), or within limits be decorated (page 92), and still preserve its neutral character.

Drapes

Drapes have long been established in neutral staging. We light them to reveal their characteristic nature. Frontal lighting reduces the sculpted folds of draped materials. Edge lighting emphasises their shape – if we are not careful, to such an extent as to coarsen their effect.

The amount of light that drapes require to reveal their structure depends largely on their material and surface finish. Dark velours 'soak-up' light, and may require very emphatic modelling. Glazed drapes reflect light all too readily; so much, that we may be embarrassed by hot-spots that 'block off' and obscure detail.

Translucent drapes (e.g. ninon) require sympathetic lighting if we are to show their attractive qualities fully. Frontally illuminated, they appear flat, blank, and uninteresting. Judiciously lit from an angled ¾-back lamp, they come to life. But over-lit, they 'burn-out' in unmodelled folds. One of the most effective approaches to lighting translucent drapes is to illuminate the area beyond the drapes strongly, so that the scene shows through them, softened and transformed.

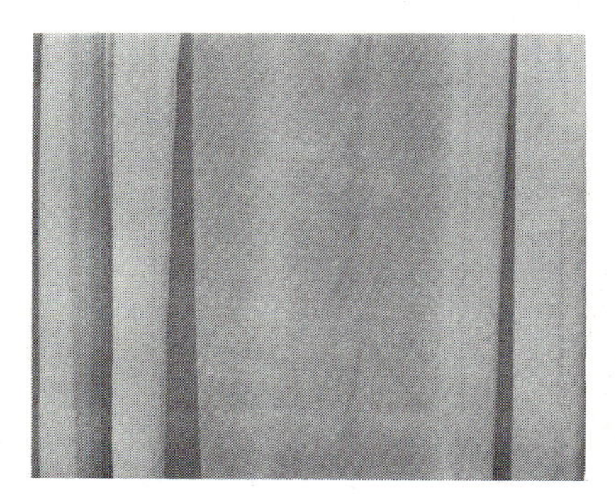

SCENIC LIGHTING – NEUTRAL SETTINGS

Lighting drapery
The appearance of drapery changes according to the angle from which it is lit. Here the same drapes have been keyed dead frontally, A, from off-centre, B, and obliquely (edge lighting), C.

Translucent drapes
Lit frontally, these ninons appear opaque. However, by strongly lighting the subject beyond and reducing frontal lighting, their translucency becomes evident as they soften subject details.

Scenic Lighting – Decorative Effects

Light offers us endless decorative opportunities. The methods are prosaic enough, but the effects one can achieve may beggar description. As mundane an idea as reflecting a spotlight from crushed kitchen foil, can create a spreading light pattern that intrigues the eye, and is a decorative delight.

Spotlight patterns

With spotlights alone, we can produce several background lighting effects. Pointed straight-on, *Fresnel spotlights* provide blobs of adjustable size, with graded soft borders. Steeper spotlights can form 'arches' on the background. At even more oblique angles, we get shafts and streaks of light. Using the barndoors, we can shape their beams into slits, columns, square or rectangular shapes.

The *projection spotlight (profile spot)* can give us hard-edged spots and light shafts. Its internal shutters enable us to shine light in four-sided patterns of various shapes. An iris attachment allows the coverage (spot size) to be varied. An internal iris may provide intensity control.

Projected patterns

Projection spotlights extend our opportunities further. Using metal stencils, we can project realistic window effects, including rose-windows, leaded lights, prison bars, etc. Designs, too, can be introduced, including stylised and geometric motifs, decorative patterns, logos, etc.

By projecting extremely defocused patterns, we can devise interestingly abstract displays. A few vari-sized holes or random slits punched into a metal mask foil, or a piece of wire mesh, or stamped-out metal sheet, can be used to produce striking light designs.

More advanced profile spotlights also enable us to project images from glass slides. These can have photographic, hand-drawn, or scratch-off designs, used singly or superimposed in intermixed forms.

There are a few practical considerations to bear in mind when using projectors. Their light output must be sufficient for our application. We usually need to reduce nearby light levels to make patterns effective (particularly when using multi-tone or finely-perforated slides). Sometimes the patterns from a series of projectors can be juxtaposed to form a larger overall effect. To avoid distortion, and to achieve overall sharpness, we must ensure that stencils are flat, (they may warp when hot), and that the projector is aligned perpendicular to the background. Projected image size depends upon the slide image size, the projector's lens angle, and the length of throw. Wider-angle projector lenses (e.g. 35°) tend to have a lower transmission, but permit shorter lamp distances.

Spotlight patterns
Standard spotlights can produce a series of simple but effective light shapes that can be used for background decoration.

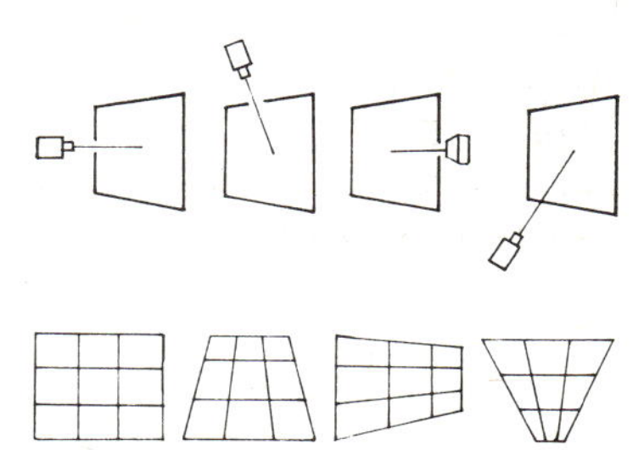

Image distortion
Only straight-on projection, at right angles to a surface, produces an undistorted image. Any angled projection inevitably results in size and sharpness variations – although these may appear quite natural.

Environmental Effect

In realistic staging, we are seeking to reproduce a particular type of environment with a certain atmospheric effect. The setting itself has been devised to achieve the appropriate locational associations, and mechanically suited to the director's production treatment, but to be fully convincing, it must be suitably lit. Lighting must convey on camera not only the physical illusions of space and structure, but enhance the style and quality of this environment, imparting a sense of occasion, conjuring a mood.

Methods of approach

There are definitely no short cuts to deriving environmental effects. They come only from sensitivity, experience, and flair. But that does not mean that the student need flounder or resort to flat, characterless illumination. Again, there are pointers to good practice.

Walls should usually be shaded to be darker towards the top. The light cut-off should not be too abrupt, and for many purposes can be roughly around shoulder height. This helps face tones to be more prominent against darkening walls. A wall containing a window to an exterior is usually darker than its neighbours.

Windows are valuable features in most settings, for they can reveal locational and temporal information. An external lamp positioned just above head height can introduce 'sunlight' into the room, casting attractive shadows over adjacent walls.

Backings outside windows may be used to simulate sky, a scenic exterior, or some architectural feature. Such backings are usually lit evenly overall, although they are sometimes decorated with shadows, for example from a tree branch. Internal backings beyond doorways (preventing cameras showing other parts of the studio when doors in settings are opened) are often shaded, and slightly brighter than the main room, so as to imply distance (page 78).

Ceilings are avoided as far as possible in studio staging, and are restricted to the essential. Ceilings make lower lamp heights inevitable and can restrict light access. So much, in fact, that we may have to position lamps to suit the production mechanics rather than the optimum pictorial treatment. Ceilings should appear neither over-bright nor unlit. Concealed lamps (e.g. on the ground behind furniture) may be used to illuminate them.

Practical lamps in the form of table-lamps, wall-fittings, should seem to influence the local brightness of settings. We invariably have to 'cheat' the effect, supplementing it with studio lamps, as the actual fittings usually prove either ineffective or too intense on camera.

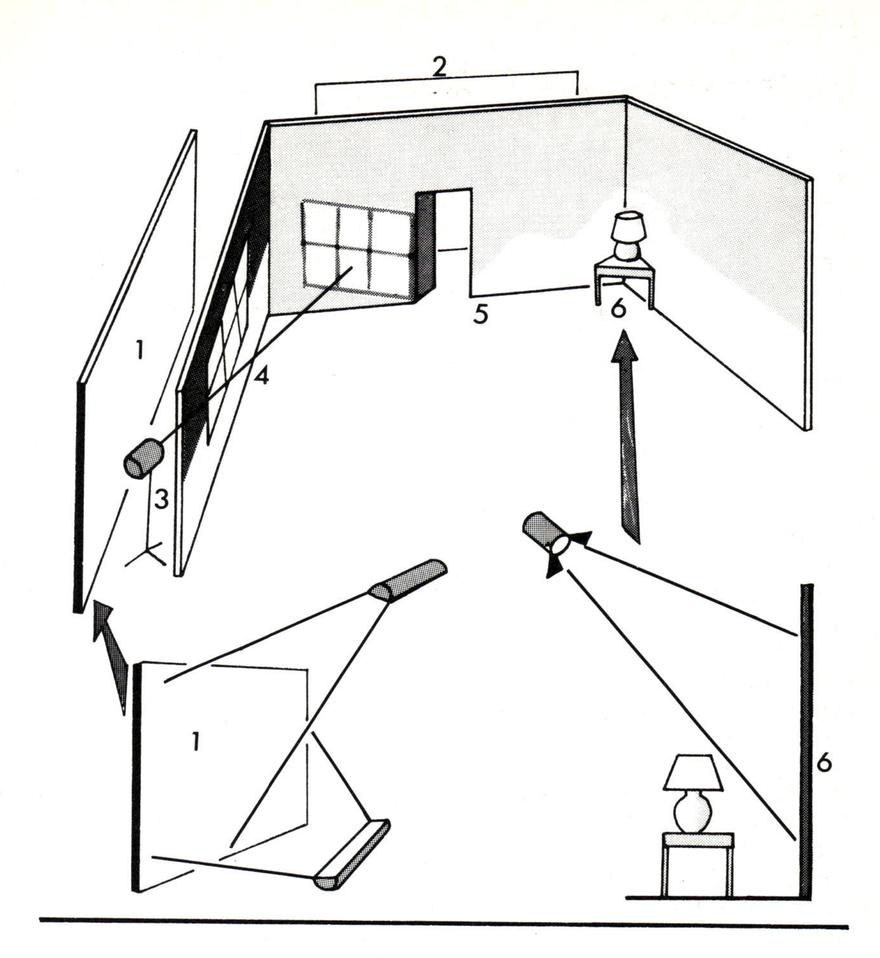

THE ESSENTIALS OF ENVIRONMENTAL LIGHTING

Certain broad principles are followed when lighting realistic settings:
1. External background (backing) – Evenly lit overall; suggesting brighter levels than in the interior.
2. Internal background (e.g. hallway backing) – Shaded and marginally brighter than the room.
3. 'Sunlight' casts light patterns over internal walls. Predominant key lighting is from the direction of the windows.
4. Walls containing windows are kept slightly darker.
5. Room walls are normally shaded around shoulder height.
6. Practical lamps seem to produce local illumination.

Faults in Scenic Lighting

When lighting a setting, it is easy to become so preoccupied with some aspects of the operation that we overlook others. Certain pictorial faults arise regularly. We should look out for them and remove them as a routine. Listed here are various defects that we shall meet. Some are aggravated or caused by staging, others by inappropriate techniques.

Typical defects and their remedies

Tones of the staging and the set dressings may exceed the tonal range of the TV cameras. If a tone promises to be too light, shade it off. If too dark, increase light on it. But take care that adjacent areas are not incorrectly lit as a result. Overall contrast range should be around 20 : 1.

Uneven lighting can be pictorially attractive. But ensure that the darker and lighter variations appear appropriate and purposeful. Generally try to avoid bright areas near the top of the picture, with darker areas below. Where lighting is intended to be even overall, don't be content with blobby, patchy illumination.

Spurious highlights, in the form of bright hot spots or specular reflections, may prove unavoidable. First aid includes a dulling spray (antiflare) on glossy or shiny surfaces, or repositioning the lamps.

Spurious shadows of many kinds can arise. Shadows of slung lamps, practicals, or scenery can fall upon the setting and the performers. We must check for shadows or unevennness on surfaces intended to look flat or unobtrusive – e.g. sky cloths, gauzes, cycloramas. Avoid oblique lighting on these surfaces.

Ugly shadows are usually grossly distorted, intrusive, or obscuring important detail. Elongated straggling shadows, clumsy architectural shadows, or 'inexplicable' shadows that are difficult for the viewer to identify, all require attention.

Multiple shadows should be avoided; although they cannot always be eliminated, particularly where movement and multiple camera angles are involved.

Spill-light may be responsible for extraneous light streaks or shadows on nearby walls. When setting any lamp, do not overlook that it may be causing spurious effects elsewhere. Illumination from a lamp lighting a foreground subject may travel beyond it to distant parts of the scene.

Lamps in shot can be a mutual embarrassment. Avoid this by co-ordination and careful planning. Lamps slung below wall height are particularly susceptible. On very long (distant) shots, lamps have often to be rigged proportionally higher or masked off, to avoid their appearing in the picture.

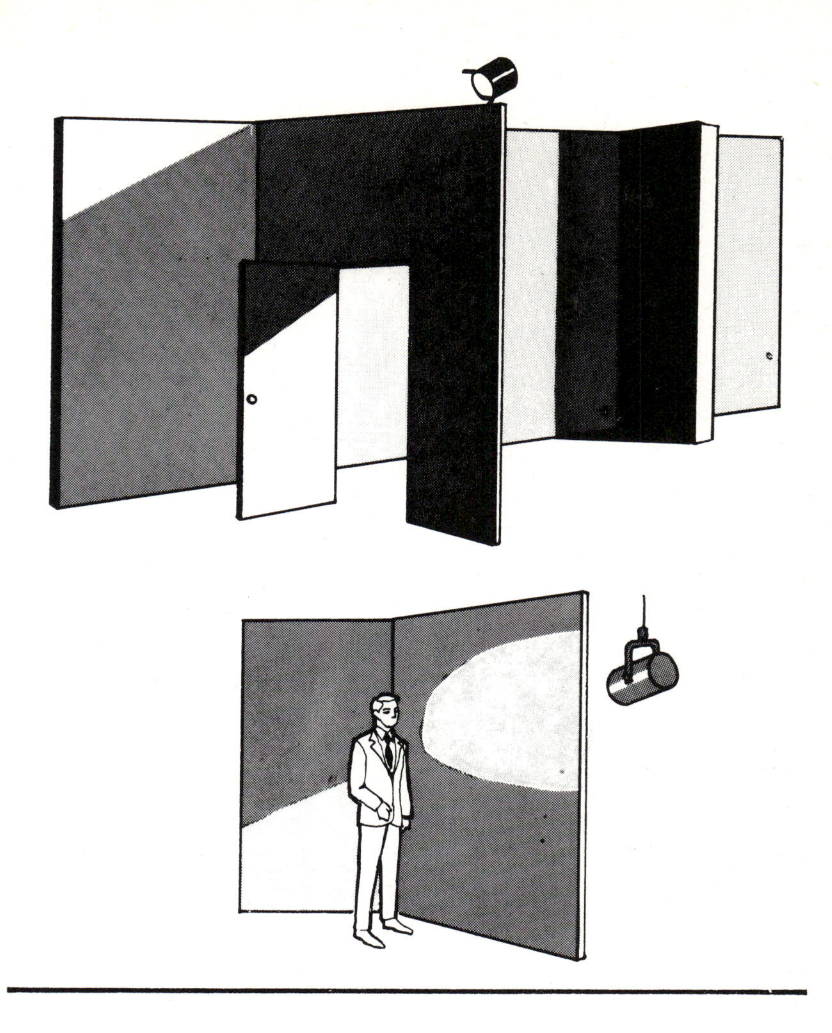

FAULTS IN SCENIC LIGHTING

Structural shadows
Shadows cast by architectural structures can create false impressions of shape and distance when seen in the camera's isolated shot.

Spill light
Take care that light from an action key light does not accidentally spill onto or 'scrape' adjacent walls, and produce distracting, unmotivated highlights.

Lighting Open Areas

When we have to cover a large floor area (e.g. over 45 sq m/55 sq yds) there are several possible approaches.

Single-key coverage
Relatively large areas can be lit with a single powerful key light. A 10 kW tungsten-halogen, or an HMI, source could be ideal here. But to be successful, this lamp must usually be at least 4.5 m/15 ft high, and around 6 m/20 ft from the action centre. Otherwise, subjects nearer the lamps are liable to be over-lit, while others distant from it remain insufficiently illuminated. Paradoxically, with the steeper distant key, there is now the prospect that its vertical angle will be too steep for closer subjects (page 106), although the light level is now constant overall. (The only real remedy is an involved system of graded diffusers.)

Dual keys
Here we subdivide the open area, keeping overlapping coverage (and hence multiple shadows) to a minimum. Where action is located mainly either side of the staging area, *opposite dual keys* can be effective, supplemented by central fill-lighting. More controllable, is the technique of *adjacent keys*, in which two juxtaposed keys are used. Their light beams are butted together (using barndoors) so that they combine to cover a wide angle. Their backlights may similarly conjoin. The effect is three-point lighting overall.

Sectional keys
Another approach to large-area lighting is to form a continuing pattern of cross key lights and three-quarter backlights. The drawback is that the spotlights' beams diverge to produce overlapping wedges of coverage. Some 'doubling-up' is, therefore, inevitable. Nevertheless, this technique can be a successful one where multi-floor-shadowing or multiple modelling in overlapping regions does not obtrude.

Soft light may be 'layered' to fill successive regions in depth as shown here; remembering that the effective strength of each diffused source falls off relatively quickly with distance.

Soft frontal
Where our soft-light sources are powerful and directional, we may be able to use them in a sectionalised frontal pattern, supported by three-quarter backlights. But this method can produce unreliable and irregular results, that can prove difficult to control.

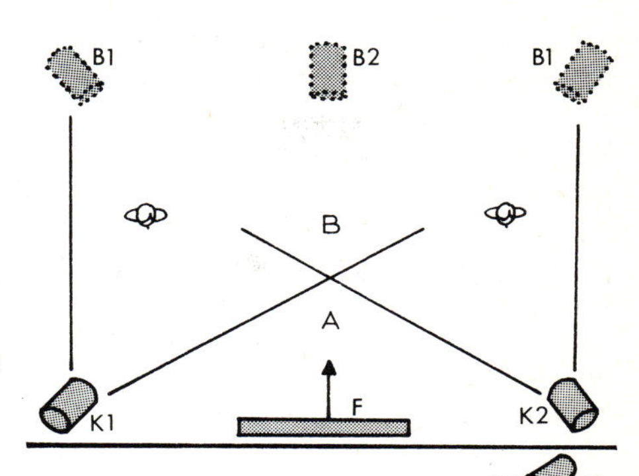

Opposite dual keys
Satisfactory for split staging, but liable to leave a large central area unlit (A) or doubly lit (B). Frontal soft light may supplement illumination from F, filling the unlit area (A).

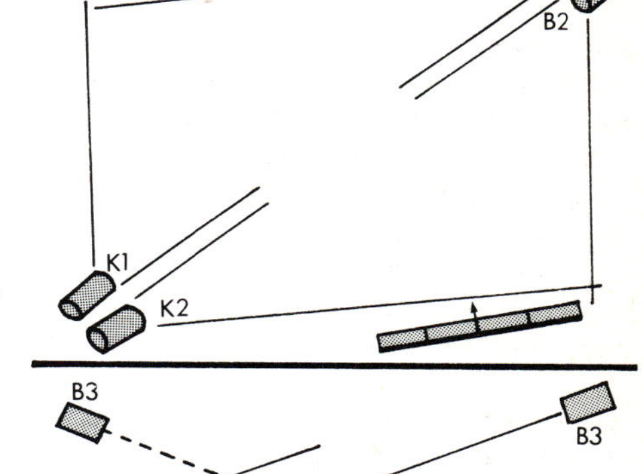

Adjacent keys
Here two adjacent keys combine to cover a wide angle. Side flaps of barndoors are adjusted to butt their beams together.

Sectional keys
Each key lights a wedge of the area, supported by a balanced backlight. Spaced rows of soft lights provide filler.

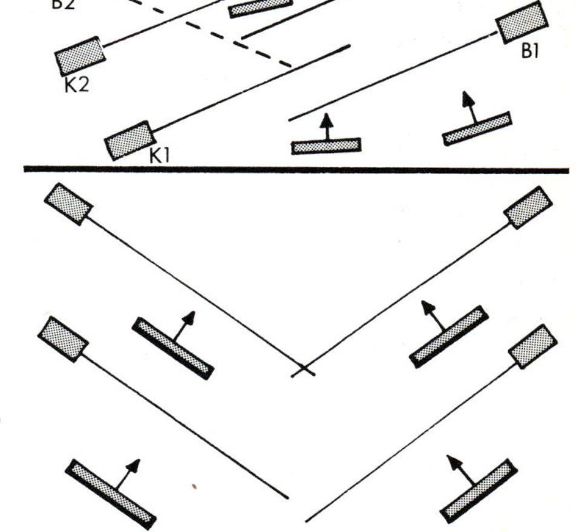

Soft frontal
Where soft-light units are powerful and directional, they may be used to sectionalise the area, modelling being provided by a series of ¾-backlights.

Lighting in Confined Spaces

In the process of TV lighting we encounter two kinds of confined spaces: when shooting in cramped quarters, and when lighting access to staging is restricted.

Cramped quarters

Both in real locations and in studio reconstructions, one encounters tiny claustrophobic situations, where there is just about room for the camera to peer in, and little else! Closets, cupboards under stairs, booths, 'phone boxes, tents, elevators, all pose the problem of confined space. We have to discover where we can arrange lighting without its being in shot, or the result being inappropriate.

A camera light, in the form of a flood or internal-reflector lamp fastened to the top of the camera head is a useful accessory in confined spaces, but resist any temptation to swamp the action with such light. Frontal lighting is often environmentally unsuitable for restricted areas, and reflects as hot spots on backgrounds. Moreover, it may over-light close subjects. Spotlights on cameras have the additional disadvantage that their effect can often be seen to change as the camera pans around.

Sometimes one can rig a lamp above the camera-access opening (camera trap), or direct a spotlight past the camera, barndooring it tightly to avoid camera shadows.

Small lamps may be attached to structures near the camera, or to either side of the action. Screwed on, stuck (*gaffer taped*), clamped, or clipped miniature light units, are invaluable for this purpose. Stand lamps get in the way, and are too easily knocked. Top light produces stark results – for dramatic purposes only. Rather than use top lighting, introduce side lighting to enhance modelling, with careful shuttering (or flags) to prevent unattractive light streaks on the background. Light break-up with a cookie, can prevent light direction from being too blatant – especially where, in reality, no illumination should exist.

Restricted access

Occasions arise where we need strategic background lighting, but have little space for orthodox approaches. We may be able to light the area sufficiently through nearby openings in the staging (e.g. gaps between flats, through open doors or windows). Light fittings can be concealed behind furniture, overhanging or jutting architectural features (e.g. buttresses), or even placed on the floor and masked with a gobo.

Safety is particularly important when working in confined spaces. Lamps emit a considerable amount of heat. They can scorch, even set fire to nearby drapes and furnishings. Cables should be concealed, and clipped up to avoid entanglement.

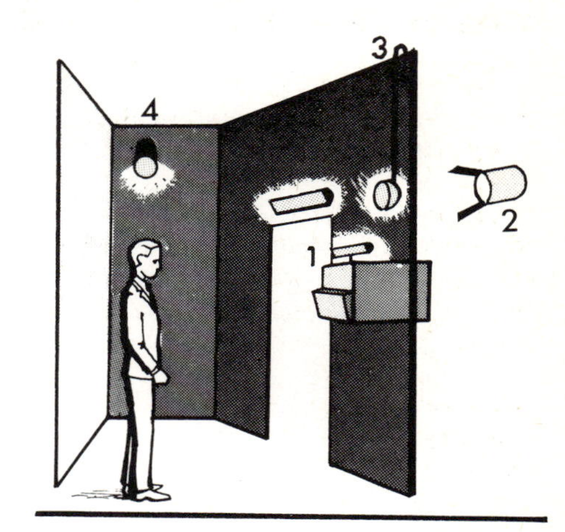

Cramped quarters
Lamps can be:
1. Affixed to the camera.
2. Shone over the camera.
3. Hung or attached above the camera access point.
4. Clipped on side walls out of the camera's shot.

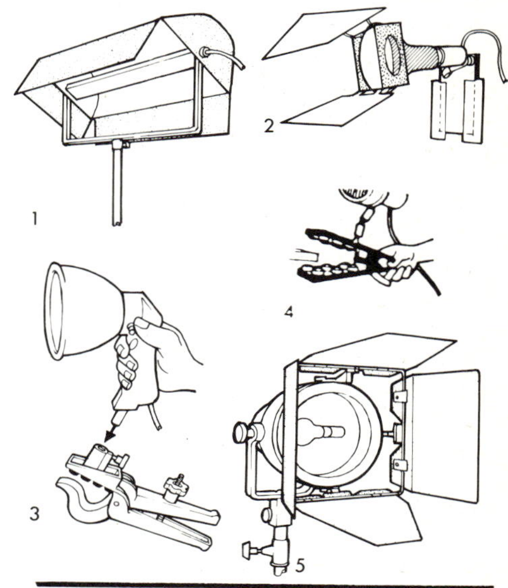

Useful light fittings
Small lightweight fittings for use in confined spaces are:
1. Small broad, nook-light.
2. Internal reflector; lamp attached to wall with gaffer tape.
3. Hand-held lamp; fits into screw-clamp (gaffergrip).
4. Sprung alligator clip.
5. Lightweight lensless spot (external reflector).

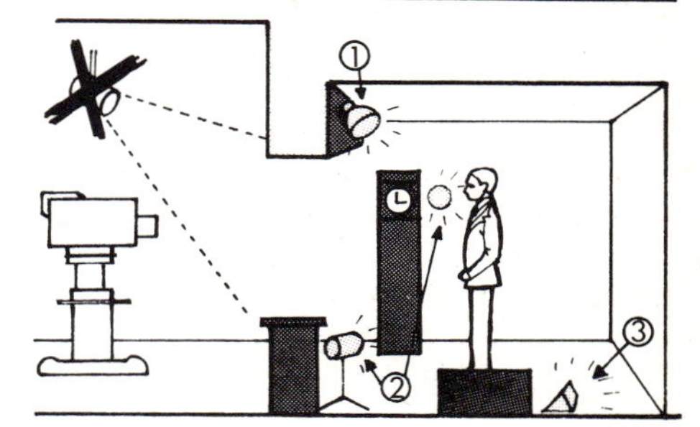

Restricted access
Where regular studio lights cannot reach the action, lamps can be concealed in various ways:
1. Behind architecture.
2. Behind furniture.
3. On the floor.

Lighting Cycloramas

Most TV studios have among their regular staging units, that maid-of-all-work, the *cyclorama* (cyc). Formed from stretched cloth (e.g. duck, canvas) and curved in a shallow 'C', the cyc may be a permanent or stored facility. A plain, single-tone area, the cyc is used variously in off-white, light grey, black, and light and dark blue finishes.

This chameleon surface offers unparalleled staging diversity – from 'walls of a room', to sky (with projected clouds, perhaps), to 'infinite space'. And most of these transformations are achievable by lighting alone.

The plain cyc

We can use Fresnel spotlights to light a cyclorama, merging their soft edges to obtain a uniform brightness overall. But this can require experience to achieve good results, and adjustment is time consuming. If the cyc is high (e.g. over 3.5 m/12 ft), two sets of spotlights may be required, to light the upper and lower regions respectively.

General-purpose soft-light sources can be used effectively to light the cyclorama, although one cannot always merge their outputs smoothly, or prevent overlapping.

Wide-angle hard-light sources incorporating parabolic reflectors offer spread illumination to cover a large cyclorama. But again, it can be difficult to obtain even brightness overall, or prevent their spilling onto other areas.

Special *cyc-lights* have been designed to enable cycloramas to be lit from relatively close distances. These tungsten-halogen strip lights are so constructed as to reflect hard light in an elongated lobe, which spreads remarkably evenly over the cyclorama surface.

A row of cyc-lights can be positioned as a *ground row* to illuminate in an upwards direction, to produce an attractive gradated effect. The lamps are often hidden behind a low vertical scenic plane, a ramp, or a concave strip *(merging cove, ground cove).* A combination of hung cyc-lights and a ground row enables us to illuminate the background evenly overall.

The decorative cyc

Sometimes set designers attach decorative motifs to the cyclorama to give it a new look, but for the most part its appearance is altered by lighting variations (page 92). A particularly valuable characteristic of lighting, is the ease with which it can transform the pictorial effect, even while the camera watches (page 116). Just by cross-fading lamps, we can take the cyc from unlit black, through graded ground-row lighting, to a bright overall state. Patterns, shafts, patches of light, can appear and disappear on cue. Little wonder that the cyclorama is ubiquitous!

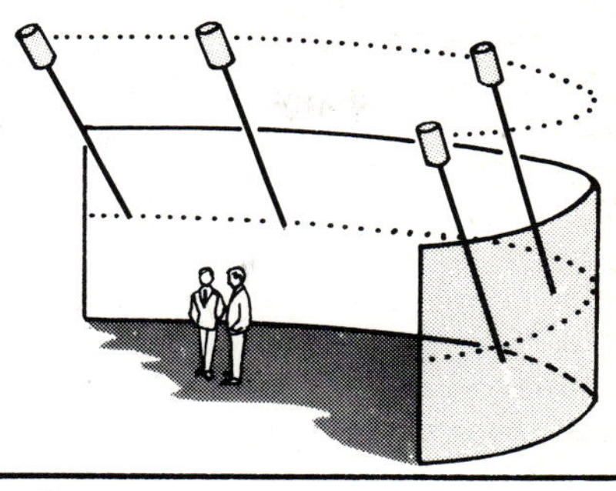

General illumination
Cycloramas (cycs) are used widely as neutral staging surfaces. To achieve even illumination, lamps should be equidistant from the cyc. If distances vary, differences in light spread and intensity will result.

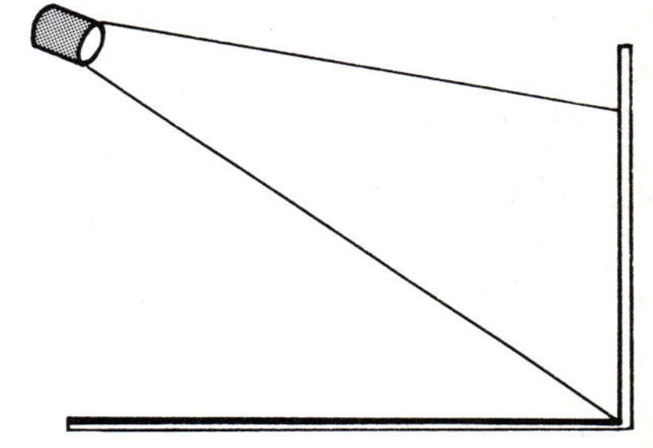

Spotlights illumination
Fresnel spotlights are useful for cyc lighting, where we want shading, blobs, shafts, or patches. Even lighting over a large area is feasible, by merging a series of soft-edged Fresnel spotlights.

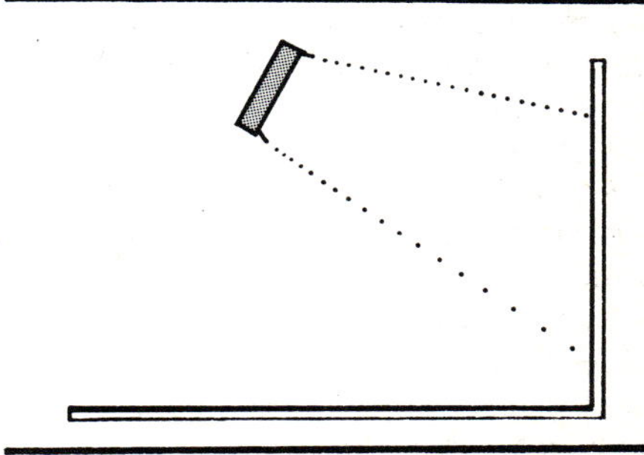

Soft light illumination
Soft light illumination falls off severely with distance. So when hung soft light units alone are used to illuminate a cyc, vertical shading can result, with an overbright upper part, and falling brightness towards the bottom.

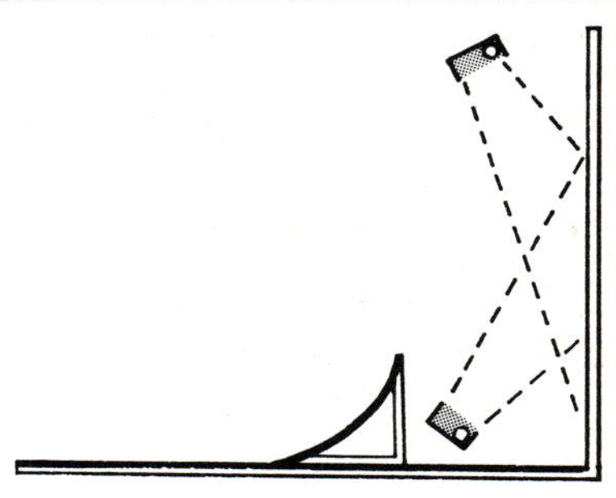

Cyc units
Compact cyc units are designed to reflect light of even intensity over a large area from an oblique slung or floor position. Hidden behind a ground cove, the upward lighting can be graded from light to dark over the cyc surface.

Lighting Translucent Backgrounds

Translucent backgrounds formed from stretched sheets of cloth or plastics, provide interestingly versatile staging devices. They have extensive applications: as screens for rear projection (page 114), to display patterns and other decorative effects, as frosted panels providing backings to fretted motifs, etc.

Methods of lighting

Lit from the front, a translucent surface will appear to be a normal, flat, usually undecorated plane. It is only when it is lit from behind, and frontal illumination reduced, that its particular magic becomes apparent.

If we do the obvious, and rear-light a translucent background from a central position, the probability is that the camera will see the lamp as a pronounced *hot-spot* that changes position as the camera moves around.

Even illumination usually requires a distant light source above the screen height, or two side lamps cross-lighting the translucent background from the rear.

A row of ground lamps will illuminate the screen with an attractively gradated effect – lighter at the bottom, falling away in intensity towards the top. But we have to take care to avoid a disturbingly bright, burnt-out base to the background.

Shadow effects

Shadows and light patterns can be cast on to the rear of translucent screens by any of the methods we discussed earlier (pages 80, 82). Cut-out stencils placed before a compact hard light source are effective, and for maximum sharpness the shapes should be set up quite close to the screen.

More than one shadow can be projected at a time, but if we use several light sources, we have to take care that the spill-light from one source does not dilute shadows created by the other. If one lamp is brighter than its neighbour, we can arrange for their admixed shadows to have varying depths of shade.

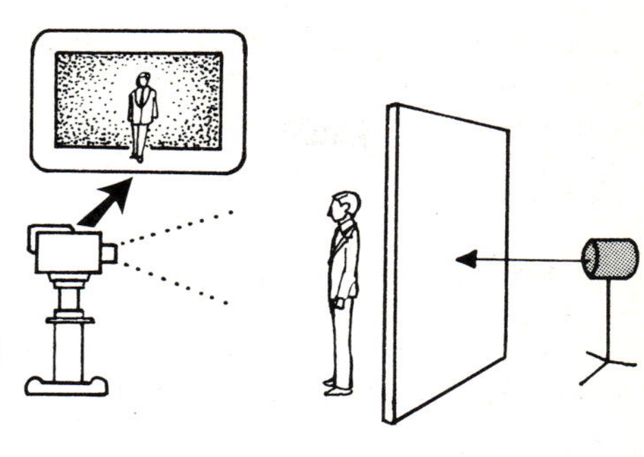

Central lighting
Central rear lighting can produce uneven illumination; a central bright area falls off in intensity to dimmer edges.

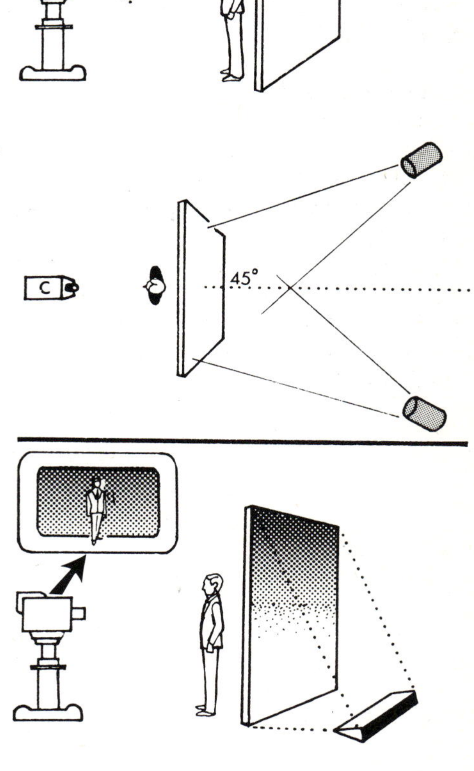

Even lighting
Even overall lighting is essential for certain applications. It can be achieved by a single lamp or by twin sources.

Graded lighting
A ground row provides attractively graded background brightness.

Lighting Multi-height Staging

Most action within the TV studio takes place at floor level, but for certain types of production elevated areas are necessary; and these bring their own particular problems.

Rostra (parallels)
Demountable platforms of various heights are used in TV staging; from a slight step-up of a few centimetres, to raised areas a couple of metres or so high (e.g. over 6½ ft). As well as simulating architectural features, these rostrum (parallel) units provide better layout and display, and more varied camera shots for choirs, orchestras, audiences, fashion shows, etc. Stepped rostra are constructed from a series of conjoined units of progressive height.

We may be able to use the same key light to illuminate both the people at ground level, and those standing on the rostra. However, if the rostra are high, we have to use a powerful lamp quite far away to achieve overall coverage. Then difficulties can arise if shadows fall onto the background behind the rostra. If, for example, actors' shadows are seen on a cyclorama intended to represent the sky, the illusion is ruined. Raising the communal key light to lower the shadows, is then liable to degrade portraiture for people at floor level.

It is better, in these circumstances, to arrange two sets of key lights instead; one for the lower action, and the other for performers on the raised areas.

Backlight positioning, too, requires care. People on the rostra, being closer to hung lamps than those standing on the ground, can become over-lit by a lamp intended to light people at floor level.

Stairs
Staircases are another common staging feature where anticipatory lighting brings its rewards. Where a long staircase is involved, this may be lit best by two or more localised keys. If one lamp alone is used to cover it, the lighting angle becomes progressively steeper towards the foot of the stairs.

In arranging backlight for the staircase, it will usually require side shuttering (to avoid light streak down the wall), and top shuttering (to prevent a lens flare for a camera looking up the stairs). Another lamp may be added, side-lighting the action (preferably barndoored to a diagonal slit following the line of the stairs), but must be well confined to maintain the effect of enclosure if the stairs are intended to be in a hallway.

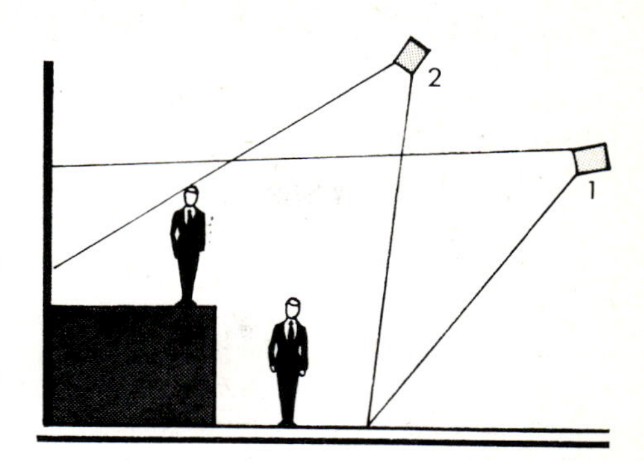

Single key coverage

1. A communal key light may prove suitable for ground level performers, yet too shallow for those on the rostra.
2. Raised to overcome this, its angle may steepen unduly for ground level action.

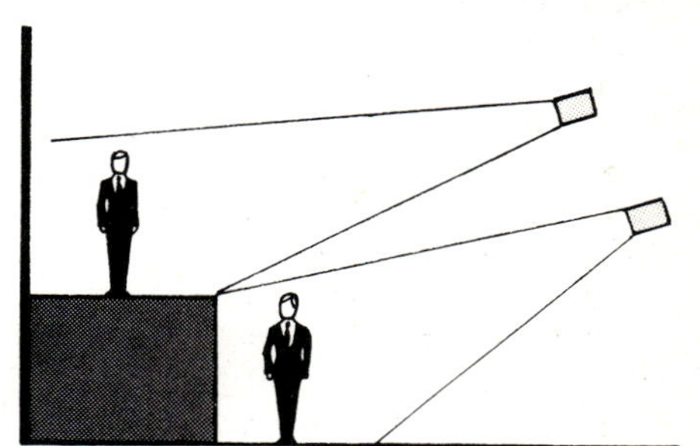

Separate key light

Action at floor and rostra levels may be lit separately. But avoid:

1. Distracting background shadows, or
2. The backlight for floor subjects overlighting the rostra subjects.

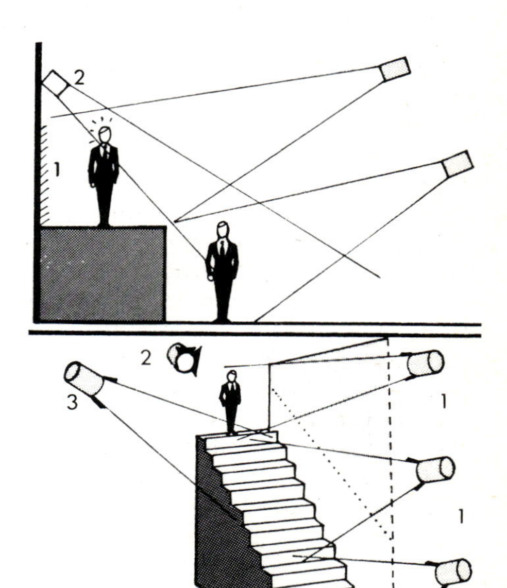

Stairs

The staircase lighting has been subdivided:

1. Individual keys light sections of the route.
2. The backlight is shuttered.
3. The side light follows the line of the stairs.

Using a Following Spotlight

The *following spotlight* is essentially intended for display. It isolates with light. It draws audience attention to a selected circular area. The spotlight can be used as the *sole* illumination, leaving its surroundings in darkness; or we can use it to provide a *brightened region* within a lit scene.

The true following spotlight is specially designed to follow action smoothly and with precision. But for slight improvement and for static situations, you can use profile spotlight (effects projector, projection spotlight) or, at a pinch, even a Fresnel spotlight with an attached *snoot* (page 50).

The following spotlight's design

The following spotlight is a familiar adjunct to stage presentations, and large area spectacles. Its light source is typically a tungsten-halogen lamp, (e.g. 3000 W), or a metal-halide (HMI, CSI) source. Despite its bulk, the assembly is balanced to enable it to be swung around rapidly and easily, with considerable precision. Controls can vary the spot size, intensity, and edge-sharpness.

Operating the following spotlight

As the performer moves around, the spot operator keeps the circle of light centred. As the action changes, he may broaden the beam to include widespread hand movements, or restrict it to encompass as little as a single hand.

This much is obvious. We have all seen it many times. But there are operational points that are less apparent. If the spotlight has a very long throw (e.g. 12 to 18 m/40 to 60 ft), as in a theatre, for example, even a slight angular movement of the spotlight is magnified considerably by the time the light beam reaches the distant stage. So it requires fine judgment and a steady hand, if the operator is continuously to locate his light beam accurately. Nothing looks sloppier than a long shot in which the spotlight misses, or partly illuminates its subject, striving to keep up with it!

When a spot has to be lit on cue, and hit its subject immediately (no nervous searching or repositioning when he discovers it is off target!), this is done either by sighting, or by pre-checking the spot position surreptitiously beforehand, while audience attention is elsewhere, with a dimmed small-aperture setting.

Whenever following spotlights are used, we must ensure that space through which their beam passes is clear of any scenery or hanging lamps. Otherwise we are likely to find shadows marring the operation.

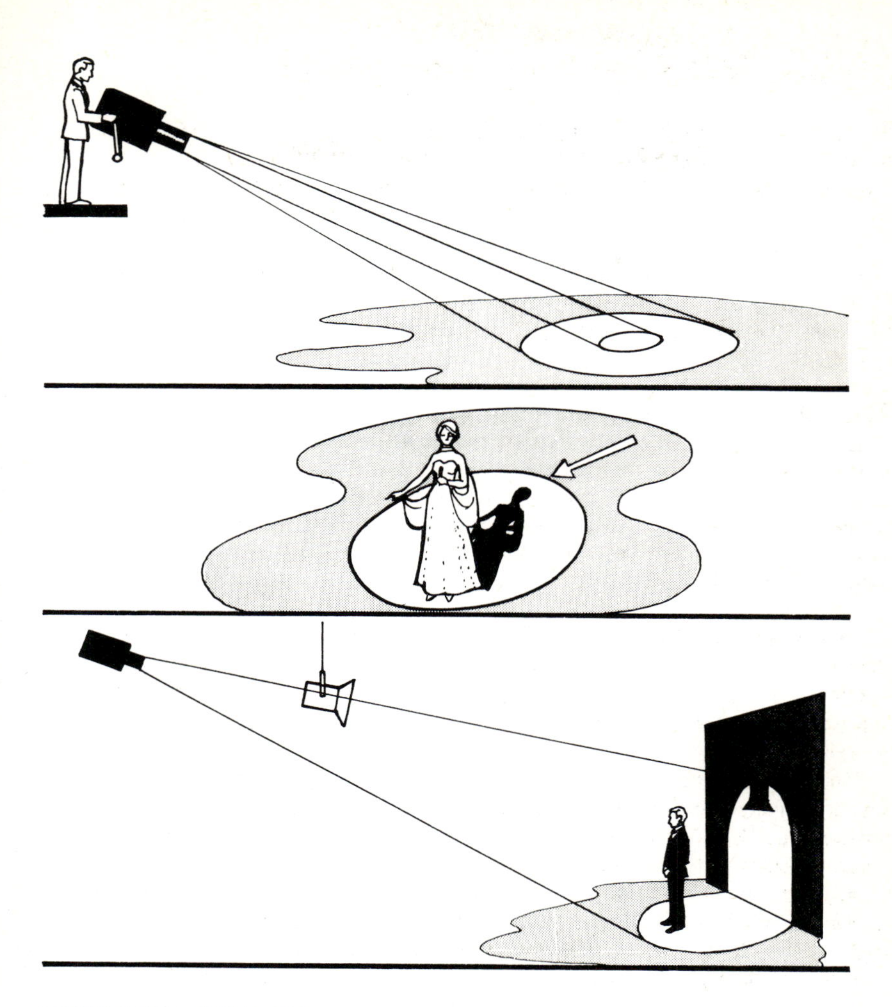

USING A FOLLOWING SPOTLIGHT

The spotlight
The diameter of the spotlight beam can be adjusted to suit the subject. It will vary with the subject's size and distance, and with the coverage required.

Positioning the spot
The easiest way to check the spotlight's coverage is to examine the position of the subject's shadow within the disc of light on the floor. In this example we can see that the safety margin is too narrow. Slight subject or spotlight movement, and it will no longer cover the subject.

Obstructions
Take care to ensure that there are no shadowing obstructions within the spotlight's field of action.

Don't forget to take spares (lamps, fuses, etc.).

Location Lighting – Interiors

Location lighting practice usually lacks the opportunities and facilities of the TV studio. Improvisation and compromise are more or less unavoidable, so that individuals tend to develop their own preferred techniques. However, the principles involved are indentical to those we have already examined (pages 36, 54-66, 72, 98, 100).

The project

Where the interior is spacious, and only a handful of lamps available, there are several possible approaches. We can restrict the camera viewpoint. Surroundings may be masked-off to confine the action area (e.g. using canvas screens). Sometimes we can shoot the interior sectionally in a series of separately-lit video-taped takes, that are later edited together. But whenever the director requires fully-lit large-area shots, there is little for it but to hire the full quota of equipment the subject needs.

In smaller interiors Fresnel spotlights are often too narrow-angled to spread sufficiently in the distance available. Instead, broader-beam sources are used. High-intensity compact grouped units (cluster, mini-brute) can provide valuable wide-angle luminants. Lightweight, high-intensity sources using high-efficiency tungsten-halogen lamps in open reflectors have proved both practical and adaptable for general purpose location interiors.

Ingenuity will enable us to get maximum value from limited resources. A single lamp's light distribution over the scene can be modified by attached flags (solid) or diffusers (scrims). Overall soft light can be produced (albeit very inefficiently) by reflecting *bounce-light* from a strongly-lit ceiling or adjacent wall. (The reflected light's colour quality may become too impure for colour systems.) Care is needed with reflected light, for we are equally likely to produce mysterious streaks and light patches over the scene, from any nearby mirrors, glass, and metalwork!

Existing illumination

Very occasionally, the existing illumination will meet our camera's needs; in intensity, contrast and direction. We may be able to position the subject to suit prevailing light, perhaps using lamps to augment it. Existing location lighting arrangements can sometimes be uprated by replacing their light bulbs with our own high-intensity versions (e.g. Photofloods).

On the whole, prevailing light is unreliable. Daylight alters in strength, direction, and quality. Existing artificial light is invariably inappropriate for our purpose (steep, wrong direction, unsuitable instensity or quality). It is prudent, therefore, to be organised to light the scene from scratch if necessary, and to take into account probable variations in lighting conditions due to time and weather.

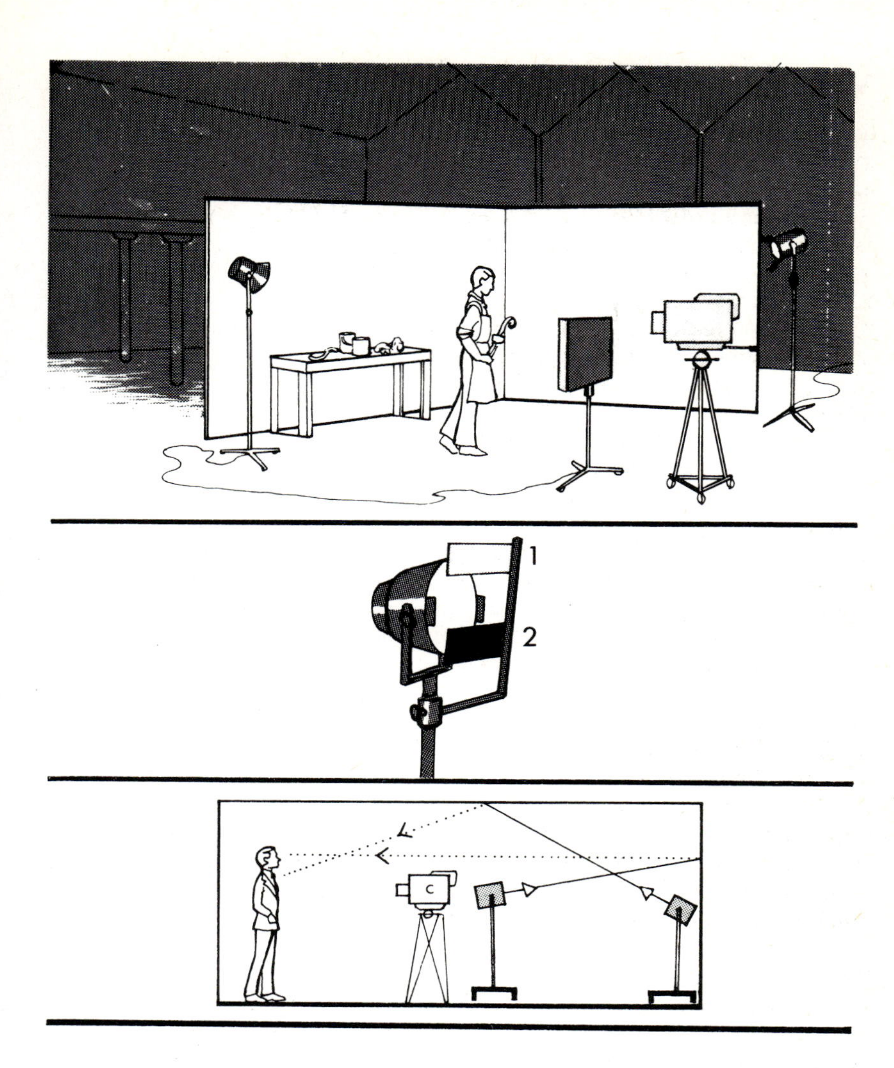

LIGHTING INTERIOR LOCATIONS

Confining coverage
By confining the area seen by the camera, the action can be lit more economically.

Controlling light
We may spread the light from a single lamp over a fairly large area, and control its local brightness and coverage by scrims, 1, and flags, 2.

Bounce light
Light can be bounced either from the ceiling or walls (or reflectors) to provide soft fill-light.

Location Lighting – Exteriors

Unaugmented daylight is a very 'take-it-or-leave-it' commodity! We have to accept it as it happens to be, or alter our shot to suit the prevailing light direction. Failing that, we have no choice but to wait for more appropriate light conditions, or try to manufacture our own.

Natural light

In TV, we can usually compensate for overall brightness changes by selecting a suitable lens aperture as we continually monitor the picture (page 132). Even differences in colour quality can be allowed for by readjusting the camera channel's colour balance.

However, there are factors for which there is no rapid remedy. The *quality* of natural light can vary quickly, too. Sunlight creates strong modelling (not always appropriate) and relatively high contrasts. Cloudy skies result in diffused, flat lighting that gives little or no definition to subjects, leaving them lacking clarity and form. Throughout much of the year, the sun's *vertical angle* is considerably steeper than we would choose for portraiture, (40° to 90°). The *horizontal* light direction also alters continually, and covers only an approximately east-west orientation.

Consequently, when using daylight, we have always to be prepared to adapt our tactics to prevailing conditions. Pictorially, the attractiveness of a daylit scene is least for overcast conditions, low when the sun is directly behind the camera (frontal), effective from a side position (usually a dynamic effect), and extremely attractive when shining towards the camera as backlight (but planes facing the camera may prove under-lit).

Compensatory lighting

If natural lighting is unsatisfactory, there are only a limited number of things we can do about it. We can accept an overcast sky, and for small localised areas introduce strong key light to provide modelling. But compared with daylight, even powerful arc lights appear puny, save for restricted treatment; particularly with a stopped-down lens. When the sun shines, we can orientate action so that it becomes side or back light, and use reflected light or lamps to illuminate areas in shade.

Reflectors are simply large boards held up to reflect sunlight onto our subject. The light quality is determined by their surface finish, being hard from metal-foil surfacing, and soft from white uneven finishes. Unfortunately, reflectors need to be clumsily large to have appreciable coverage, and so are susceptible to wind rock. They are, moreover, entirely reliant upon the sunlight for their efficacy.

Portable lighting units using nickel cadmium batteries can augment close shots, but otherwise power lines from generators or public supplies become necessary for high-intensity equipment (page 134).

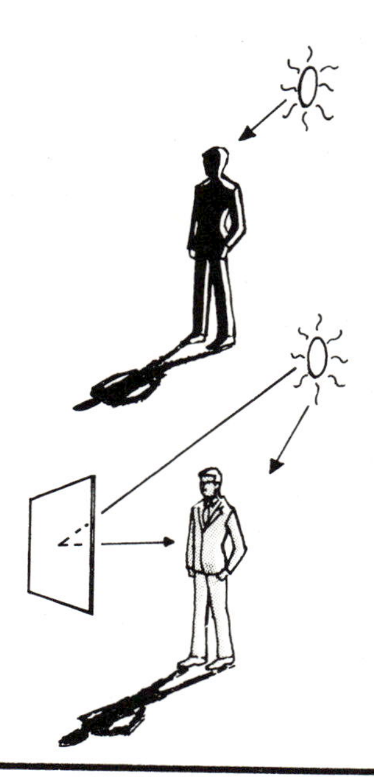

EXTERIOR LOCATIONS

Natural light
Prevailing light may not be lighting the scene from the direction we require. The lighting contrast may be excessive.

Reflectors
Portable surfaces placed to reflect sunlight can provide effects ranging from low-intensity filler, to a full key light, depending on their surface material and finish, and the prevailing light conditions.

The subject lighting must match the background scene.

Scenic Projection and Insertion

TV has several ingenious staging devices that find increasing use in studio production techniques. These have the very practical advantage of providing 'instant backgrounds' of many kinds by comparatively simple means.

Scenic projection

In *rear projection (back projection, BP)*, an image of the scene is projected onto the rear of a translucent screen. The camera shoots action from the reverse side.

We have to take care to avoid light falling on the screen, casting shadows and diluting the projected image. So the foreground subject should be kept sufficiently distant from it (e.g. 2 m/6½ ft away) for lighting to be shaded off the screen.

In *reflex projection* (front axial projection) the subject is placed before a special glass-beaded screen. The required background image is projected along the camera lens-axis (Pepper's ghost principle, page 120) onto this screen. The TV camera sees the foreground subject as if located in the background scene. Thanks to the screen's highly-reflective, extremely narrow return-angle, the image is very bright, and not susceptible to spill-light dilution. The projected scene is not visible on the brightly-lit performers, and their own images hide their shadows.

Scenic insertion

This system enables us to insert into the background picture (from any video source), a subject located in front of the master TV camera. In *monochrome* TV, circuitry for picture insertion utilises a physical mask or electronic switching pattern (keyed-insertion, inlay), or is triggered by tonal differences between the subject and its black or white background tone (brightness separation overlay).

In addition to using keyed-insertion methods, *colour* TV systems use *chroma key* (colour separation overlay, CSO). The subject is placed before a coloured background (usually blue or yellow), which triggers a video-switching circuit to insert the subject itself into the picture from any video source (another camera's shot, slides, film, video tape, etc.).

We must always ensure that the foreground subject does not include or reflect the tone or hue of its special background (which is even overall), or spurious breakthrough or edge-ragging will result.

Matching

In all of the systems, we must take care to match the foreground subject to its artificially juxtaposed background picture. Consequently, we need to check such details as scale, light direction, relative intensities, contrast, tonal and colour quality, to prevent any discontinuity or disparity between them.

Rear projection (back projection, BP)
The camera sees the subject backed by a projected picture. Spill light from foreground illumination must be minimised.

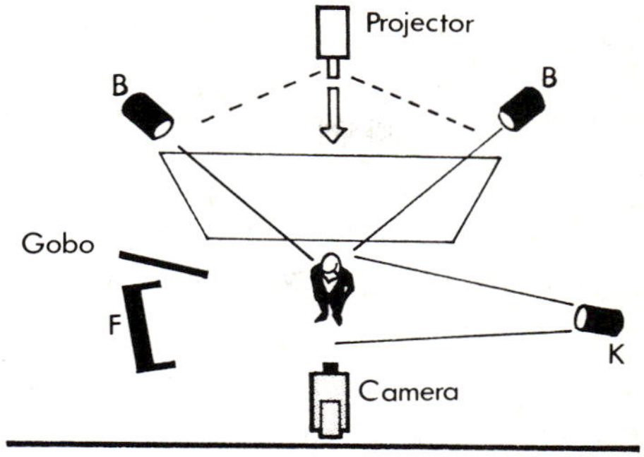

Reflex projection
The camera sees both the subject and the bright reflected image of the background scene. Foreground lighting also falls upon the screen, but is not reflected towards the camera.

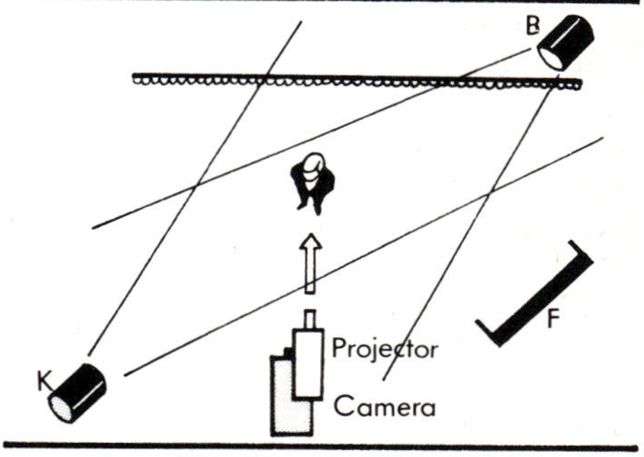

Chroma key (colour separation overlay)
The subject is placed in front of a blue (or yellow) background. Wherever the electronics detect this colour in the subject camera, 1, they switch to another video source (Cam 2).

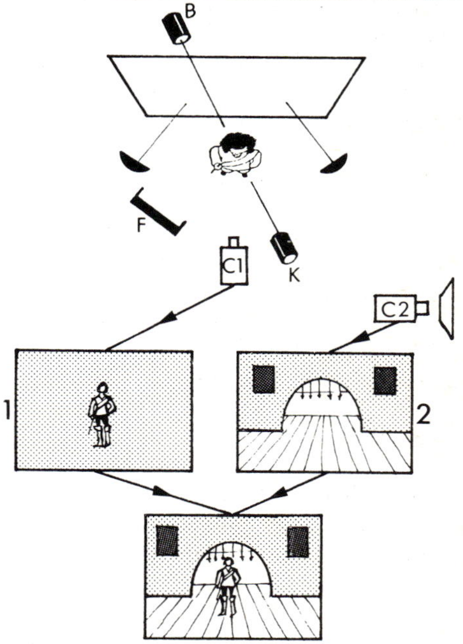

Lighting Changes

One of the most intriguing aspects of light is the ease with which we can modify the effect it creates. Our lighting produces a particular illusion. We alter the treatment, and the picture is transformed. The fundamental mechanics of lighting changes are themselves simple enough. Individual lamps or groups of lamps are switched, or faded up and down on cue. But the skills lie in the subtlety and effectiveness of these changes. How elaborately, and how smoothly they can be introduced, will depend upon the lighting control facility we have available.

Control facilities

The simplest forms of electrical control are *contactors* (electro-magnetic relays) for switching, and resistance dimmers (e.g. assembled in transportable *dimmer trucks*) for fading light levels or adjusting preset intensities. The most complex lighting control consoles provide multiple fading and cross-fading, simultaneous group switching, memorised intensity and switching relationships, etc. But for many productions, this degree of elaboration is never required.

Pictorial changes

Lighting changes are used in diverse ways in TV presentations. The actual methods do not have to be elaborate. They may involve as little as a single lamp being switched. But the result can enhance pictorial appeal enormously.

Presentational changes of several types are used regularly. For example, a person is announced, and a spotlight stabs the darkness to reveal him. At the end of a discussion, a lighting change fades the key lights, leaving the speakers silhouetted against a lit background. Again, we can move attention from one area to another, by fading the lights on the first, and bringing up lights on the new one.

Decorative changes are a regular feature of 'light entertainment' (vaudeville, band-shows, song-and-dance shows). As the mood of a song changes, one decorative display slowly alters to another. We might emphasise musical beat by rhythmical light-switching. By introducing a lighting change, we can denote the end of a song, or herald the next act, e.g. lights up, following a low-key item.

Environmental changes are found in TV drama, where atmospheric lighting is changed instantly. Commonest, perhaps, are situations in which people enter darkened rooms and switch on lights.

Temporal changes are used occasionally in TV drama, e.g. the room illumination altering very gradually, as night falls.

BACKGROUND PATTERN CHANGES

Background patterns modify how we react to the foreground subject. By changing background lighting alone, we can alter the subject's prominence and significance.

Using Coloured Light

Colour provides snares for the unwary! Coloured light offers great opportunities. But just as easily, it can accidentally produce garish or bizarre results far from those we intended.

Using colour media

To obtain coloured light, we use special thin plastics or gelatine sheeting. This colour medium, available in an extensive range of hues, is fitted taut in metal frames clipped in front of lamps. The 'colour frame' permits air circulation and so prolongs the useful life and colour of the medium. Never fasten medium directly over the lens. The material will crisp up, and the lens (or bulb) may break through overheating. Gelatin sheeting is quite cheap, but is brittle, buckles, and discolours easily. Although thin plastics sheeting (e.g. Cinemoid) is more expensive, it has greater durability and hue constancy, and is non-flammable.

We should judge a filter's effect by shining light through it, rather than holding it up to a light, as this can prove deceptive. The exact result any colour medium produces is influenced by the lamp's brightness, its colour temperature, and the surface colour it illuminates. In a colour picture a white surface can look multi-hued (orange/yellow/blue) if lit with 'white light' of varying colour temperatures, although it may appear of quite even intensity in monochrome reproduction.

Coloured light on people

People are seldom convincingly lit by coloured light, except for startling effects. Completely unnatural colours are the most striking, but even less audacious orange or yellow illumination (used for firelight, candlelight, etc.) quickly palls. Only white light really produces attractive portraiture. The low colour temperature light from dimmed lamps can produce unacceptable colouration in the form of red-orange faces or yellowish hair-light.

Coloured light on backgrounds

Coloured light has great potentialities when creating attractive background display effects; whether for a naturalistic illusion (e.g. blue sky, sunset), or a decorative one (e.g. suiting the mood of music).

Be cautious about using strongly-coloured backgrounds behind faces. Because skin tone tends to be subjectively altered by the colour of its background (moving towards the complementary hue of the background) this can modify the apparent skin colour in the final image.

The effect of colour temperature
As the lamp's colour temperature increases, the light becomes more bluish. Lowering the lamp's 'kelvins' results in a warmer colour quality.

Simultaneous contrast
Background tone (and hue) can affect our assessment of foreground subjects. Here the same light grey shapes are used throughout, but their changing backgrounds cause them to look lighter or darker than they are.

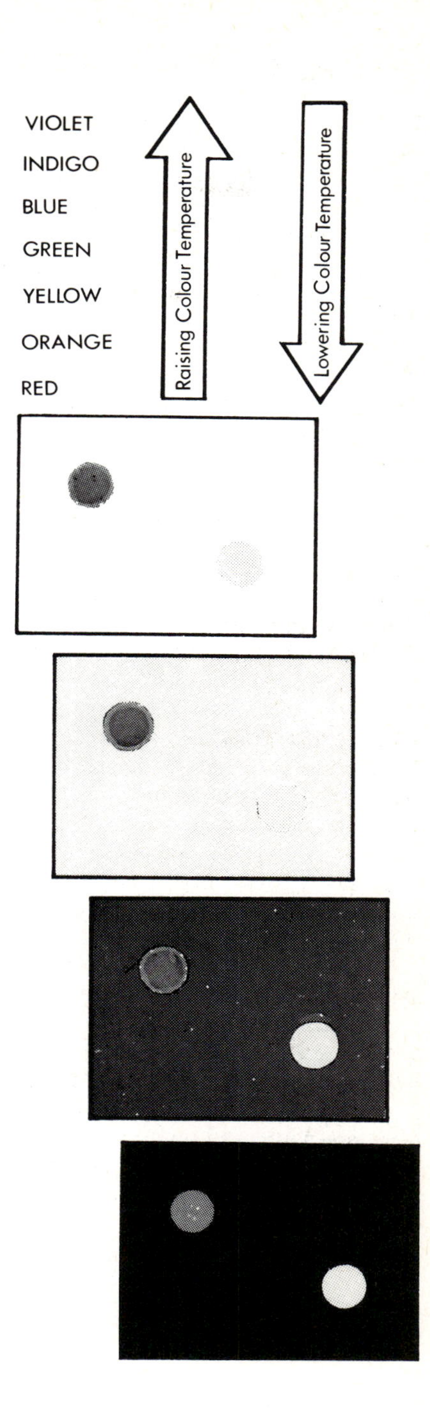

Lighting Graphics

Graphics (captions) include the various flat displays such as photographs, charts, maps, used to illustrate TV programmes. *Titling (title-cards)* too, are set up in front of the camera, to provide identification names, subtitles, cast lists, and similar written information. Normally they present few difficulties, but where the lighting is inappropriate, spurious effects can mar their appearance or legibility.

Lighting set-ups

If the lamp illuminating a graphic is set at too oblique an angle, any surface irregularities are emphasised. Paper may not be equally firmly attached to its background card, or the backing may buckle in the heat. So too, any textural variations (roughness/smoothness) in the graphics treatment, or any retouching alterations, dents, or blemishes, can become pronounced under such edge lighting. Oblique lamp positions also cause horizontal or vertical light fall-off (shading, sawtoothing) across the caption.

Even illumination is essential for the best results. A soft light either side of the graphic is a good general solution. Failing that, a Fresnel spotlight some 3 m/10 ft or so away, at a 20° to 30° vertical angle (not above 2H) usually suffices.

A shallow-angle caption lamp can create a camera shadow, or cause hot spots. Glossy-surfaced captions are particularly liable to disfiguring, obliterating flares and specular reflections. Glass or clear plastics surfacing invariably pose problems for these reasons.

Pepper's ghost

By illuminating our graphics indirectly via a clear glass sheet, we can virtually project the light along the lens axis. Theoretically, the nearer the light source is to the centre line of the lens, the less the camera can perceive any cast shadows. So, by using a *Pepper's ghost* arrangement, we obtain shadowless lighting. (Soft-light sources are seldom sufficiently diffuse to achieve this.) Particularly where the graphics are multi-layer (i.e. with flaps, pull-out sections, etc.), we can by this means avoid edge shadows or overlap shadowing from these layers.

As only a low proportion of the light is reflected from the clear glass, a powerful lamp is necessary. Furthermore, it is essential for the glass to be thoroughly cleaned, to avoid any marks appearing in the picture.

Lighting set-up
A. For optimum results a soft light fitting either side of the camera is preferable.
B. A slung spotlight may be used (to keep the floor area clear), but this has disadvantages too.

Camera light
Captions are sometimes lit with camera lights. A peripheral frame ('garland') gives even illumination.

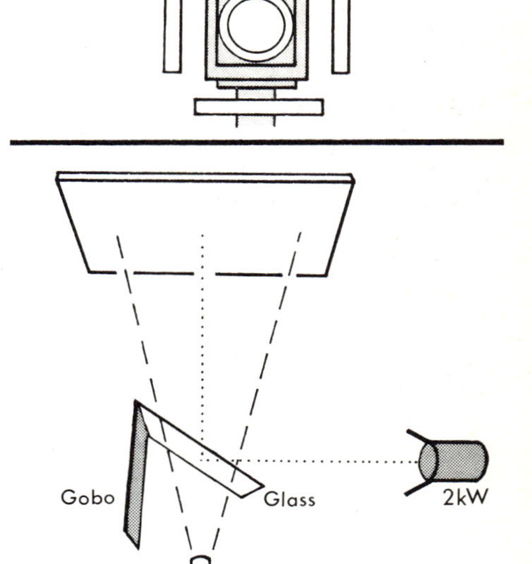

Pepper's ghost
Light is reflected from a glass sheet at 45° to the camera lens axis. The gobo prevents light spilling beyond.

Plotting Lamps

A *lighting plot* is a scale plan, showing the lighting treatment for a production or presentation. Symbols identify the different types of lamp usage, and indicate the equipment involved. Plots have several practical uses: when planning, organising and estimating lighting treatment; to provide a record of the final rig.

Although there are now standard lamp symbols (*CIE), others are widely used. These are often augmented by detailed information on rigging (e.g. drop-arms, clamps, etc.), diffusers, colour media, patching, cable routing, lamp accessories, stands, etc.

Preparatory plots

For studios where the work-load is high, or for large-scale productions, advanced planning of lighting treatment becomes essential in order to make full use of equipment, studio-time, manpower, etc., and to co-ordinate the work effort of various other production services (scenic transport and erection, floor painting, set dressing). Fully-detailed plots are prepared, and the required equipment rigged (overnight, perhaps) before scenic erection, and in readiness for lamp setting (page 124).

Where such complex arrangements are not necessary, as in smaller studios, we may be able to draw up a plot and rig simultaneously. For regular daily shows, a 'permanent' rig is often installed, augmented where necessary to suit particular action.

Record plots

During rehearsal, the lighting plot serves us as a reminder of the lighting treatment, and aids rapid identification of lamps (pages 20, 22). After the production, this record of the set-up can be retained for the file (for training purposes, to indicate equipment and power usage, etc.) and if the production form is to be repeated on some future occasion, it forms a *standard plot* to guide others.

Lighting plots are essentially only indications of our treatment. They are not three-dimensional. They cannot represent exactly the effect the lighting has achieved. We may show the lamp's height beside its symbol, but its precise angle and coverage will not be depicted. Nor will we usually record the lighting balance (dimmer/fader settings). Nevertheless, plots are an invaluable discipline, both in the way they encourage us methodically to create our lighting treatment, and for reference during rehearsal.

* Commission Internationale de L'Eclairage.

LIGHTING SYMBOLS	CIE Symbols	Others used
FLOODLIGHT (100°-180° beam angle)		
SPECIAL FLOODLIGHT (beam less than 100°)		1 2 3
SOFT LIGHT		
REFLECTOR SPOTLIGHT		
LENS SPOTLIGHT		
FRESNEL SPOTLIGHT		
PROFILE SPOTLIGHT (hard-edged beam)		
EFFECTS SPOTLIGHT (projecting slides, stencil)		
SEALED BEAM LAMP		
DUAL SOURCE LAMP (HARD/SOFT SWITCHING)		Hard Soft
CYC LIGHTS STRIP LIGHTS		

Footnote 1. Scoop 2. Double broad 3. Broad

APPROACHES

There are two fundamental approaches to lighting:

1. Choosing a lamp (or its position) by on-the-spot observation, or
2. By calculation on scaled plans, indicating by symbols the complete lighting treatment.

The table shows many commonly-used symbols.

Setting Lamps

The lamps are now fixed in position. The time has come to adjust their coverage and intensities to fit our planned concepts.

Typical approaches
Ideally, we switch on each lamp in succession, adjusting it to blend progressively, lamp by lamp, to build up the complete effect. If all the unset lighting is switched on together, accurate judgment is frustrated by their random overlapping coverage, and the overall uncontrolled glare.

It is usually best to tackle the portrait lighting, and the background lighting (environmental lighting, effects lights) separately – although circumstances may necessitate some lamps having dual roles (page 88).

Setting lamps
There are several methods of setting lamps. The obvious one is to adjust it yourself to cover the area you require. The disadvantage is that you really need to stand *where the camera will be*, to judge a lamp's effect. Many lamps (e.g. backlight) can be appraised accurately only from the camera viewpoint.

To set portrait keys, you can position yourself appropriately, with your back to the source, while someone adjusts the spotlight focus. Watch your floor shadow, while the fully-spotted lamp is centred on your head. Then flood the lamp.

Alternatively, *using a dark viewing filter*, look into the flooded spotlight, having it moved to locate the bright filament image in the centre of the Fresnel lens. Now the lamp is centred on your head. Barndoor adjustment can be determined similarly, by looking into the lamp through a filter. But as far as possible, check light by examining resultant shadows, rather than by staring into lamps. Otherwise, apart from potential eye damage, you will find your visual judgment impaired by strong after-images, and be unable to appraise tonal values accurately.

Barndoor adjustment for background lighting is best done by standing well back, examining the light spread, and closing respective barndoors to restrict the illumination to the required area – avoiding overlapping light beams ('doubling'), unwanted coverage, or undesirable shadow formations.

Point soft-light units broadside to the area being illuminated, tipping them downwards to prevent widespread spill over the entire scene.

Checking lamps
If several lamps are alight, we can check the coverage of each (e.g. to trace which one is overbright, or causing a shadow) by watching the moving shadow of a hand or pole waved in front of it, or by 'flashing' the lamp (switching it on and off).

Building up the background lighting
These shots from a TV production show how lamps have been added progressively, so that step-by-step, lighting builds up to the total effect.

Identifying lamps
For quick lamp identification, hold out a finger and join its shadow to the finger-tip. This imaginary line points to the lamp responsible. Here we see that as well as the required key, 1, we have spill-light from another lamp, 2.

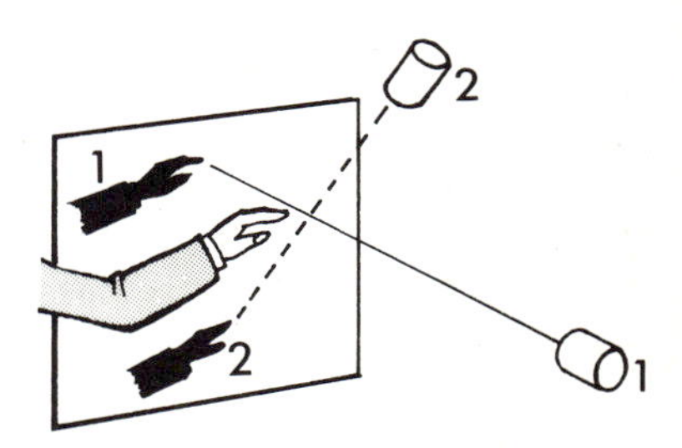

Camera Rehearsal

In television, the camera rehearsal forms an essential part of the lighting process, for it is then we check and correct our treatment.

Suitability of lighting

We design our lighting to suit certain anticipated *action, shots* and *camera positions*. If any of these changes, this lighting may no longer be entirely suitable. We may have lit for a dramatic profile, only to find that the director is now taking a full-face shot instead. So we have to alter the lighting to suit the new situation – re-angling the lamp, rigging a new one, or re-balancing existing ones are possible remedies.

Correcting shortcomings

Throughout rehearsal, we keep a critical eye on our picture monitors. The *transmission (line) monitor* shows the studio's output, i.e. the shots selected by the director, that our audience sees. A series of *channel (preview) monitors* continually show pictures from their respective cameras. Using these, we can detect any lighting defects such as spurious shadows, inappropriate balance, or unsuitable light directions.

We may correct shortcomings as they appear, or list them for attention during a rehearsal break. Some alterations require only the rapid readjustment of a fader or a barndoor. Others can necessitate lengthy revisions. But remember, when we alter any lighting treatment, we must ensure that this revision is not going to upset any earlier sequences that took place in the same area! One can overlook this, and move a light to suit current action, only to find during the next rehearsal that previous action is now ill-lit – although the troublesome scene is now quite satisfactory.

Checking pictures

During rehearsals, the *lighting plot* serves to remind us of our treatment (page 122), and aids rapid lamp identification. We can see at a glance, the lamps lighting the area taken in by a camera's shot.

If lighting control facilities are available, it is a useful trick to switch any lamp we assume to be causing a defect, while watching the picture. The flashing should cause the hot spot, shadow, etc., to come and go. If it does not, you have isolated the wrong lamp! The more one can track down and identify problems during rehearsal, the more likely are we to cure them. Never watch and list errors without forming firm ideas about what is causing them – and their probable cure.

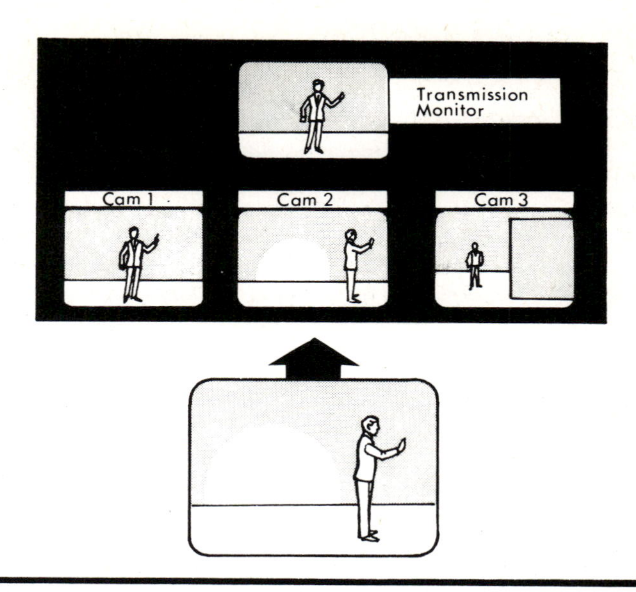

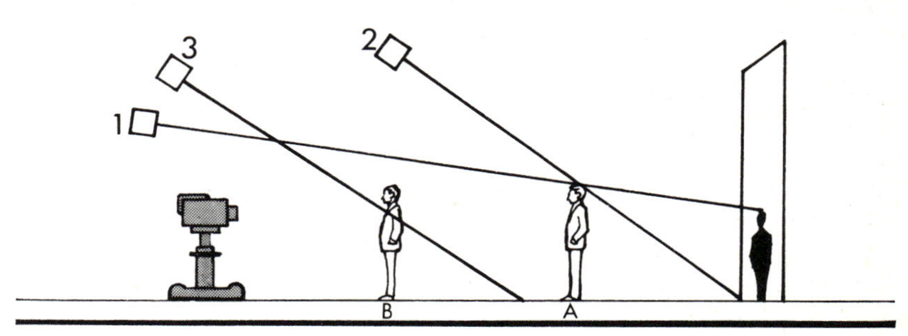

CHECKING THE LIGHTS

Using preview pictures
While Cam 1 picture is selected for 'transmission', we notice in Cam 2's preparatory shot that the cyc is unevenly lit. Taking advantage of this preview picture, we may be able to correct the error before this shot is actually taken. This 'preview check-up' technique is invaluable during unrehearsed productions, to ensure that defects are minimised before they appear on the air.

Anticipate the effect of alterations
Disliking a wall shadow behind subject A, we alter the key light from 1 to 2, moving the shadow downwards. But now, when the subject moves to B, he is no longer lit, and an extra key, 3, becomes necessary. Whenever we make an alteration, we must try to anticipate its possible effect on the rest of the production.

Measuring Light

TV has the intrinsic advantage that if we alter a lamp, or change lighting balance, we can see the results instantaneously.

Consequently, light measurement in television studios is often only a preliminary process, and levels are largely estimated. Instead of meters, we rely mostly on the *ultimate* exposure meter – *the TV camera itself*.

Preparing correct light levels

Clearly, it would be laborious and time-consuming to begin rehearsals with light levels that are grossly outside the TV camera's needs. Too little light would necessitate adding more lamps, or replacing present fittings with others of higher power. Too much light, and we would have to do the reverse, adding diffusers, or dimming lamps. Comparative light measurement is imperative, particularly for the inexperienced.

The general light levels we require will depend upon our type of camera tube, how it is adjusted electronically, the lens aperature, and the lighting contrast we are seeking. Little wonder, then, that we find differing advice about the light levels needed for 'the TV camera'. However, some representative data is included opposite as a general guide.

Methods of measuring

Growing familiar with our particular equipment, we may estimate that a 'fully flooded 2-kW spot, 4 m/13 ft away' provides an appropriate key-light intensity. If it is a little too bright, we can diffuse or dim it; if insufficiently bright, slight 'spotting' will increase the intensity. Where a dimmer-board is used, it is a widespread practice to run all lamps at about 60% of full intensity (190 V for a 240 V supply). Lamps are brightened or dimmed as needed about this average.

In both monochrome and colour TV, we measure the *incident light* falling upon the scene, and also the *surface brightness* of faces, highlights and shadows (i.e. the contrast ratios and contrast range of subject and scene). Average reflected light at the camera position is a less useful measurement; although as a rough check it may warn us of undue discrepancies. Power consumption, too, tells us nothing about actual usage.

In colour TV, we are also concerned with the colour quality of light, so *colour temperature* measurement is necessary for consistent standards. Common practice is to align camera channels to local standards (e.g. 1600 lx/150 fc incident at 2900 K) and to adjust lighting and exposure to match these parameters.

Measuring Light

TYPICAL LIGHT LEVELS REQUIRED§

Camera tube	LIGHT LEVELS LUX	FC	Stop† *f*	Scenic contrast handled
MONOCHROME				
Image orthicon 3″ tube	300–350	28–32	5.6	20 : 1
Image orthicon 4½″ (e.g. type 5820)	500–600	46–56	5.6	20 : 1
Vidicon 1″ tube	+1600	+150	2.8	50 : 1
Vidicon 2/3″ tube	+600	+56	2.8	50 : 1
Plumbicon* 1¼″ tube	500	46	4	30 : 1
COLOUR				
Vidicon 3 × 2/3″ tubes	2000	186	2	50 : 1
Plumbicons* 3 × 1″ tubes	750	70	2.8–4	30 : 1
Plumbicons* 3 × 1¼″ tubes	1600	150	4	30 : 1

* Lead-oxide Vidicon.
† Depth of field for the I.O. at *f*5.6 compares with Plumbicon/Vidicon at *f*2.8.
§ Video adjustment may increase these requirements considerably.

LIGHT MEASUREMENT

A. Incident light levels from key, fill (base) back light checked.
B. Check subject-to-background tonal contrast, to ensure good separation (e.g. 1½ : 1 or 2 : 1 typical).
C. Scene's tonal contrast measured (should be around 20 : 1 max.).

Lighting and Sound

Sound pick-up in the TV studio is provided by microphones that are worn or held *(personal)*, supported in *stands, slung*, or carried on a *sound boom*.

How sound pick-up affects lighting
Microphones used below head-height pose no lighting problems. *Slung microphones* may cast shadows on faces and scenery, but this can invariably be avoided by careful lamp positioning and judicious barndoor adjustment.

The difficulties that arise from the use of the sound boom form a major frustration for sound and lighting specialists alike. The microphone is attached to the end of an extensible boom arm. This counterbalanced tube pivots on the central vertical column of a tripod or wheeled 'pram'. Its operator extends and swivels the boom arm to position the microphone near the performer.

The microphone is always kept outside the camera's field of view, positioned around 0.6 to 1.8m/2 to 6ft from the performer, according to the camera's length of shot. Shadows sometimes fall upon the performers or background as disconcerting moving shapes. If they do, the microphone or boom will have to be repositioned, the lighting revised, or the action modified to clear the shadow.

Controlling boom shadows
The figure opposite shows where the mic shadow would fall for any of six key light positions. The questions that have to be answered are:
a *Is the mic shadow visible in shot?* Remember, there will *always* be a mic shadow if a hard key light is used. We aim to throw it out of shot, or disguise it (e.g. hide it in background shadow).
* *Can the mic be moved?* The mic must be within a reasonably appropriate distance of the sound source to match sound perspective to the picture. Sometimes it can be raised or re-angled.
* *Can the lighting be adjusted to clear or hide the shadow?* Often a barndoor flap can be used to shade the background where the shadow appears. But obviously it is irrational to degrade pictorial quality unduly to clear such shadows. Instead, it may be necessary to alter the shot (to exclude the shadow) or use a different mic arrangement (e.g. personal mic).
* *Can lighting be planned to avoid shadows?* Normally shadows are deliberately thrown out of shot by keeping keys on the opposite side of the camera lens-axis to the boom, and by using upstage or side-wall keys for cross-shooting cameras.

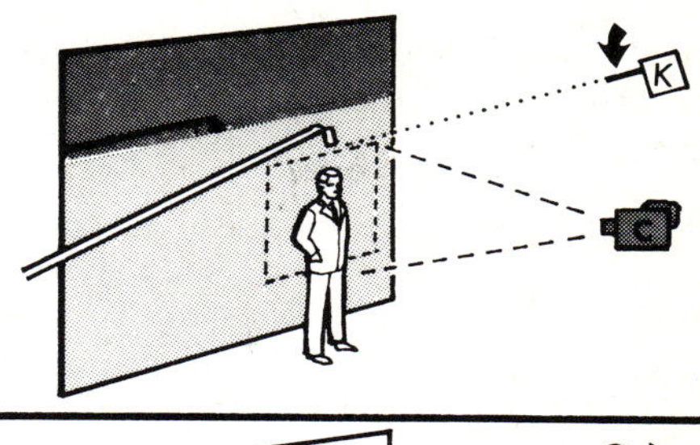

Barndoor
Lowering the top barndoor flap may shade off the area containing the shadow.

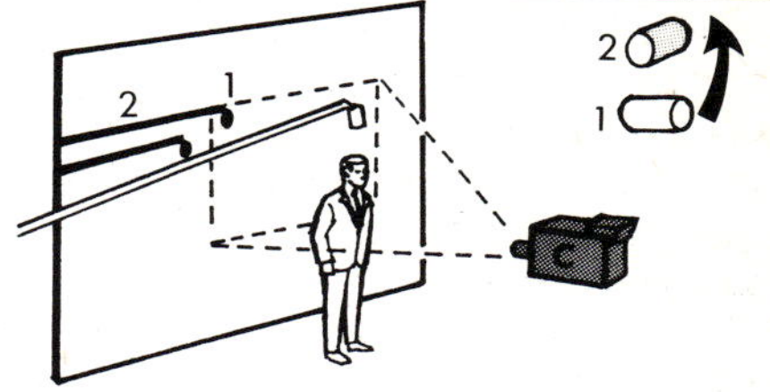

Moving the key
The shadow may be moved outside the camera's shot, by repositioning the lamp (from 1 to 2).

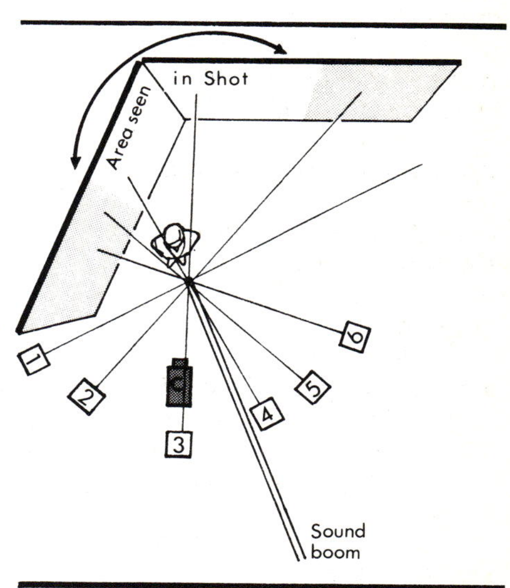

Key-light position and boom shadows
This diagram shows where shadows fall for each possible key position. Keys 1 and 2 (opposite side of camera to the boom) are typical optimum key positions for this situation. Lamps 3 and 4 would continually cause shadows. Lamps 5 and 6 avoid shadows, but are too widely angled to the subject.

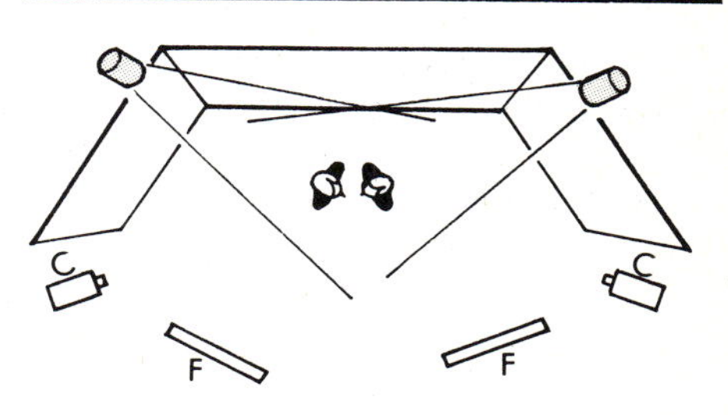

Upstage keys
Keys located in upstage positions tend to throw boom shadows downstage, and out of shot on cross-shooting cameras.

Lighting and Picture Control

It is possible to limit studio operations so that we set up the cameras' electronics, adjust their lens apertures, and then just concentrate on positioning and focusing their shots. But this can produce really satisfactory pictures only if camera movement is minimal, scenic tones restricted, and the lighting relatively even overall.

Where facilities or manpower are very limited, or where we can accept picture quality below the optimum, this 'hands-off' approach may suffice. But for the highest quality, particularly when a variety of shots is involved, each camera channel has to be monitored, and its operating conditions continually adjusted to provide consistently matched pictures.

Operational control

The *video operator (vision control operator)* has several controls available: to change the *exposure* (by lens aperture, neutral-density tone-wedge, target voltage), the *black level (sit, set up)*, the *gamma* (in monochrome cameras), and the *colour balance* (in colour camera channels).

Exposure adjustment enables him to selectively control the amount of light the camera-tube receives from the scene. The camera can reproduce only a certain limited range of tones, and by exposure adjustment, he selects which segment of scenic tones is reproduced well. He can expose to get excellent gradation in light materials, such as paper documents or snow (by stopping down to prevent their being over-exposed); but shadow areas will clog so that no detail will be seen in them. Conversely, he can expose for shadow detail (by opening up to expose lower scenic tones fully). But now very light tones will *block off (burn out)* and be seen only as blank white areas. Clearly then, the more restricted the scene's tonal range is, the easier it becomes to accommodate it and to expose to reproduce its subjects clearly.

Black level control normally adjusts the TV picture tones, so as to relate the darkest to the minimum point in the video signal (*black level*). We can 'sit up' the entire reproduced tones (greying out shadows, but producing no further shadow details – e.g. for mist effects), or 'sit them down' (crushing them towards black – e.g. for deep shadows, night effects, consolidated black backgrounds).

Gamma control relates to the harshness ('bite', 'snap', 'vigour') of the tonal reproduction, or its tonal subtlety. In monochrome, it can alter subjective quality and pictorial effect.

Techniques

Briefly, the video operator adjusts controls to make the best of each shot, giving attractive and appropriate results and matching interrelated pictures. He can also modify the pictorial quality created by lighting and staging; so close co-ordination is essential here.

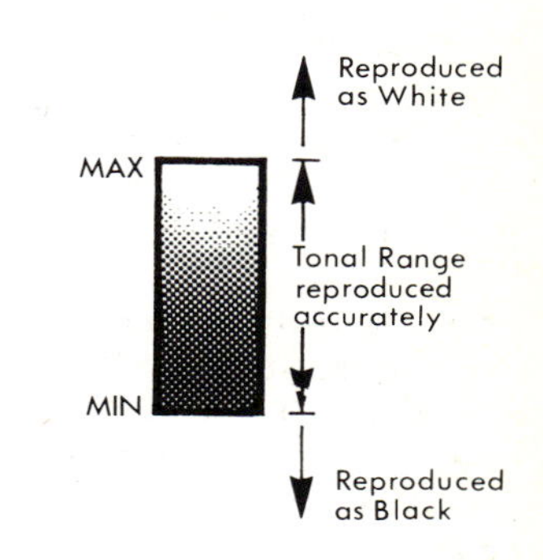

The camera's tonal limits
The TV camera is able to reproduce tones accurately only within a limited range of values (e.g. 20 : 1 or 30 : 1).

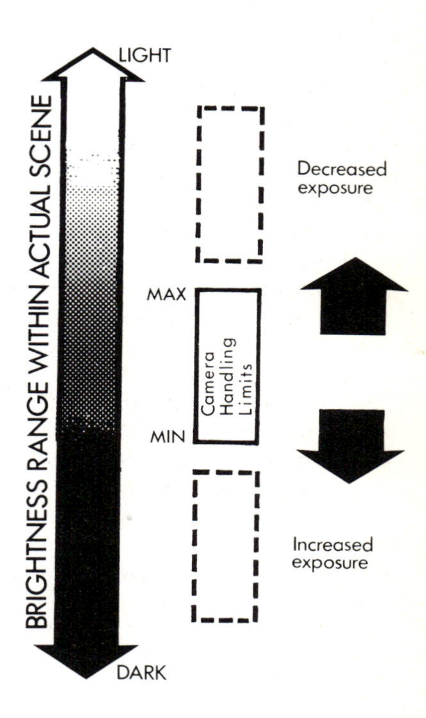

Exposure
The scene usually contains a wide brightness range (deliberately restricted in the studio).
As we adjust the camera's *exposure*, we are actually selecting which segment of the whole tonal range in the scene we wish to reproduce accurately – for the camera can handle only this limited portion.

Lighting Power Supplies

The actual provision of lighting power is a matter for the expert. Poor earthing (grounding) can cause fatal accidents. Substandard equipment, half-knowledge and 'working lash-ups' can together produce horrifying results.

Power sources

In the permanent studio, figures around 250 W/m^2 (for up to 800 lx) are quoted for lighting power. But clearly our requirements must vary with the size and elaboration of productions. We must often tailor our aspirations to the power available.

Public power supplies are locally 'standardised', so that in Europe 220 V 50 cycle supplies are common, in North America 110 to 120 V 60 cycles, and in Britain 230 to 250 V 50 cycles are in general use. Provided that our equipment is designed for operation on the supply used, the practical differences are of only peripheral concern to most users. There may be 'phase' considerations, which preclude equipment fed from different power phases being used in close proximity. There may be the need to have d.c. supplies for certain equipment (e.g. arcs, camera dollies), as well as the normal a.c. supplies for studio lamps.

Household power supplies are necessarily very restricted (e.g. 13 or 30 A maximum per circuit), but can be adequate for small areas, when using overrun lamps. Even portable battery supplies may suffice under these conditions.

Field equipment usually involves hiring portable generators or alternators (130 to 200 A typical capacity using gasoline or diesel fuels) where high voltage, industrial or similar regular supplies cannot be utilised. Unregulated power sources are load-conscious, their voltage dropping as power consumption increases (this results in progressively lower light outputs from each lamp, and falling colour temperature). Great care is needed when load-shedding (switching off), as voltages can suddenly rise and burn out any lamps remaining lit. (Mechanical governors on generators or *automatic voltage control* can alleviate this problem.)

Cabling

The normal basic procedure in routing power to mobile lamps, is to run a main cable (feeder-line) from the supply source, to a multiple *junction box (splitters, spider)*, and then to feed individual lamps from this fused distribution point. Where a lamp's lead is itself not long enough, an extension cable (or 'cord') is introduced as a link. The *interfaces* (plugs and receptacles) are usually designed to prevent inadvertent overloading of cables and fittings.

Practices vary, but both fusible links and cut-outs are used as safety devices. Fuses may be included in each lamp, in its plug, or in its supply.

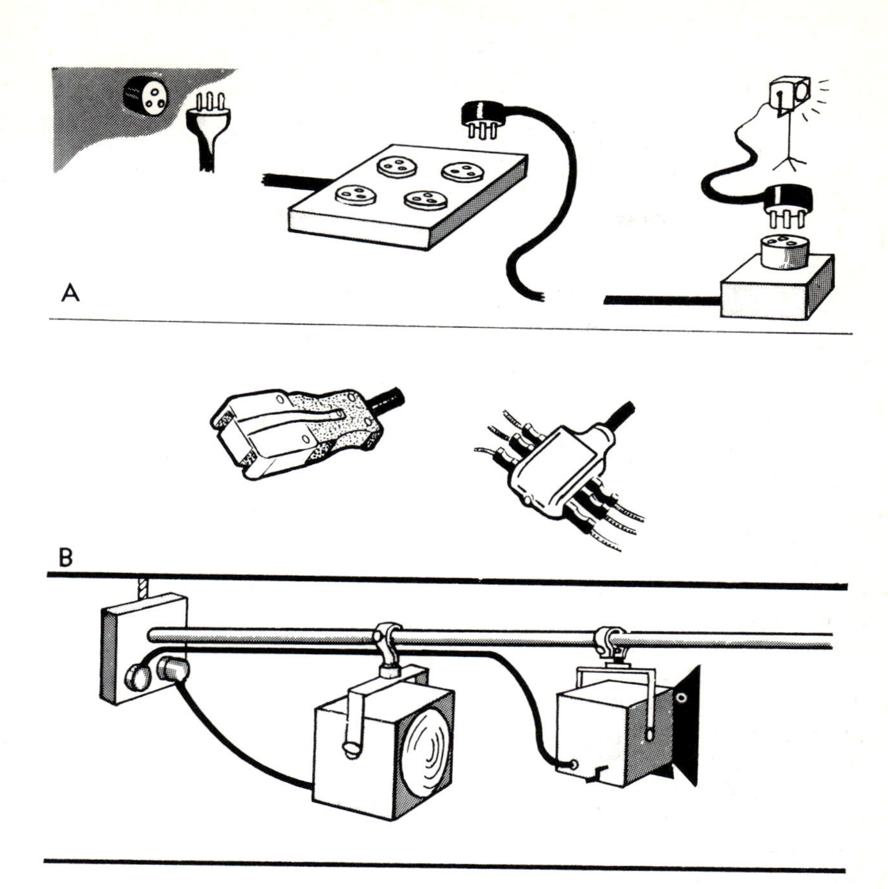

POWER SUPPLIES

Feeding the lamp
A. The main feed cable from the supply socket provides a junction point (spider) from which individual lamps may be fed. Extension cords enable lamps to be used further from main junctions.
B. An alternative, widely-used form of plug and receptacle arrangement (Kleigl stage plug and distribution box).

Local plugging
Where lamps are permanently rigged, a shorter cable ('tail') plugs into an adjacent socket.

Safety

Preoccupation and time-pressure can cause us to overlook the continual need for safety precautions. Small studios easily become congested. In large studios, people are busy in their separate pursuits, and liable to be unaware of changing hazards around them. Below are reminders of the care that is so essential for the safety of all.

Lamps should have safety bonds (steel cord) preventing the fittings or their accessories from falling. Wire mesh guards should protect people from bulb explosions or lens fragmentation. Lamps must not be located against drapes, cycs, staging, etc. (fire risk). They need good ventilation.

Cables should be removed from floors; or where this is unavoidable, concealed or covered. Cables should not be run near hot lamps or apparatus. They should be clear of water or moisture. Knots and friction should be avoided. Keep cables from sharp edges or possible run-over. Cable ends should be firmly interconnected (tied together, if necessary, to prevent pull-out). Cables should be used only within their rated limits.

Lamp stands should be heavily bottom-weighted if raised. Stands can easily become top-heavy. Ensure that their cables are not taut, and are locally secured.

Lamp heights can be so low that people or apparatus can foul them. If this cannot be avoided, ensure that clear warning flags mark them.

Hot fittings are a hazard, even when handled with asbestos gloves. Moreover, hot lamps are fragile.

Lamp movement must be carried out with care. Lamps can be lowered inadvertently onto people or furniture, or swing out of position if brail lines slip.

Ladders, steps, treads must be positioned firmly when in use, with someone holding them fast. They should be stored securely, not left against the wall.

Earthing (grounding) is imperative for all electrical equipment – lighting fittings, practical lamps, electrical apparatus, staging metalwork, and supplies.

Water must not be allowed to fall upon lamps, cables or connections. Bulbs explode if subjected to water spray, rain, etc.

Phases. Equipment fed from different supply phases must be kept well apart, to avoid any possibilities of shock from voltage differences between them.

Fuses and circuit breakers should be rated for the items they supply.

Equipment condition can easily deteriorate, from dust to cable-fraying, rust and corrosion.

Overheating can take place, with the prospects of shortened lamp life or even fire, if fittings are enclosed.

AVOID makeshift, improvised 'lash-ups' using temporary bindings round connections or damaged cable!

SAFETY

Stand lamps
When stand lamps are fully extended, they tend to become top-heavy. So weight their bases (sandbags, stage weights), and ensure that their cables cannot be pulled taut.

Low suspended lamps
Avoid suspending lamps below head height. They are liable to be moved accidentally. If a low lamp is unavoidable, hang a prominent warning.

Lamp handling
Avoid handling hot lamps. Don't work over action areas. Don't use steps that are insecure and unsupported.

Further Reading

MILLERSON, Gerald:
The Technique of Lighting for Television and Motion Pictures
Focal Press, London and Boston
The first comprehensive study of the art and techniques of lighting for television and film cameras. Suitable for student and specialist alike.

MILLERSON, Gerald:
The Technique of Television Production
Focal Press, London and Boston
A comprehensive study of television production methods, staging treatments, sound, etc., including the interrelationship of these various crafts with lighting techniques.

MILLERSON, Gerald:
Basic TV Staging
Focal Press, London and Boston
A survey of fundamental methods and techniques in television scenic design and construction, including their relationship to lighting.

NURNBERG, W:
Lighting for Photography
Focal Press, London and Boston
A study of the principles and practices of photographic lighting.

NURNBERG, W:
Lighting for Portraiture
Focal Press, London and Boston
A detailed, systematic study of the effect of light on portraiture.

SAMUELSON, David W:
Motion Picture Camera and Lighting Equipment
Focal Press, London and Boston
Includes extremely useful data on film lighting equipment; much of which also relates to television production lighting.

Glossary

acting area That part of the setting/staging area/environment, in which performances take place.
ambient light General illumi ation falling onto nearby surfaces. Particularly referring to random light spilling on a screen – e.g. room lighting illuminating a TV picture tube (receiver or monitor) and degrading the image contrast and colour quality.
anti-flare (dulling spray) (72) A fine wax coating sprayed over shiny surfaces in which specular reflections appear distracting on camera. The coating diffuses the reflection, but if overdone, may modify the appearance of the subject.
auto-transformer (44) An iron-cored coil across an a.c. supply voltage that enables a variable voltage to be tapped off to provide adjustable lamp brightness.
automatic voltage control (134) Automatic voltage regulation (AVR) ensures that variations in the power drawn from a supply do not cause changes in the supply voltage. Without such control, increasing power loads can cause the line voltage to fall; or to rise rapidly (surge) when an appreciable load is removed – possibly burning out ('blowing') remaining lamps.
available light (12) A general term for the unaugmented normal illumination present (natural and artificial) at a particular time and location. This has usually to be modified or augmented for technical or artistic reasons to produce high picture quality.

backlight (34, 36, 42, 54, 72) Light directed towards the camera (usually on the rear of a subject) to define its outline, model its edges, or reveal translucency.
back projection (BP) (114) See REAR PROJECTION.
balance (of lighting) (46) Adjustment of the relative intensities of lamps lighting a subject (key, fill, back, set lights, etc.), to achieve a particular artistic effect (atmosphere, temporal impression, mood, etc.).
barndoors (50, 72, 131) A metal fitting attached to the front of a spotlight, enabling the light-beam to be cut off by two or four hinged metal flaps.
barrel (28) End-suspended tubular bar that can be height-adjusted, using counterweights, counterbalanced, or motorised hoists. Power points affixed to barrel-ends supply hooked-on lamps.
base light (foundation light) (22, 32, 46) Overall soft light reducing lighting contrast; so avoiding exaggerating scenic tonal contrast.
basher (40) Slang term for CAMERA LIGHT.
batten A stage lighting trough (strip-light fitting), usually suspended overhead. Term for a long lamp-support bar (as BARREL).
batting down on blacks (132) See SIT. Crushing the darkest picture tones to an even black; e.g. to obliterate unevenness in a black background caption (obviating shading or light fall-off).

black level (68, 132) Technical term for the part of the video waveform representing black in the picture.
block off (90) See BLOOM.
bloom (block off, crush-out, burn-out) (10, 12, 24, 74, 90, 104, 132) When the brightest picture tones (e.g. specular reflections, highlights) reach the upper limits of the video system (the white-clipper) they merge to become a blank white area. This excessive brightness may result from the surface finish, its tone, angle, or form over-lighting, or over-exposure.
break through (114) A spurious effect sometimes encountered when electronically inserting a subject into a background scene. An area of the background scene appears 'within' the subject.
bridge (108) A tubular-scaffolding fixture or suspended gangway, used to provide an operating position for a following-spot, or for rigging lighting equipment.
brightness (luminosity) (44) One's subjective impression of the amount of light reflected from surfaces. Often very inaccurate due to psychological effects (adaptation illusions), e.g. a pure colour (highly saturated) may be wrongly interpreted as 'bright'. The term LUMINANCE is used for accurate measurements of reflected light. (In the USA *brightness* is often used to denote *luminance*.)
burn-out (10, 90, 104, 132) See BLOOM.

cameo lighting Action against a black background. See LIMBO.
camera light (28, 40) A lamp attached to the camera to provide local frontal fill-light for close-ups, or as a travelling key light.
camera trap (100) An opening arranged within scenery through which a camera can shoot (often remaining unseen by other cameras), and so increase the potential camera viewpoints. Such devices as pull-aside drapes, shutters, hinged wall-pictures, sliding panels, are used.
camera tube (10, 128) The electronic tube located within the camera-head, that generates the video (TV picture signal) from the lens' image. Types include the Image Orthicon, Vidicon, Plumbicon, Saticon.
carbon arc (22, 98, 108, 110, 134) A luminant created by a gaseous discharge between two cerium-cored carbon rods. These *trims* burn for a limited duration (e.g. 45 to 90 mins.). While arc lamps produce high-intensity light of excellent colour quality and sharpness, they require skilled operation and maintenance to achieve these properties, and to sustain an even, constant light output.
catchlights (52) See EYELIGHTS.
catwalk (walkway) (28) A confined passage in the studio ceiling, above any lighting grid, giving access to suspended lighting and staging equipment, air conditioning, hoists, etc. *Also*, used for the narrow balcony (lighting gantry) affixed high around studio walls. Here luminaires can be attached to tubular rails, as can any manned lighting (follow-spots, arcs), staging tie-ropes, slung cables, etc.
centre stage A position in the centre of the acting area.

channel (preview) monitor (126) A picture monitor continuously displaying the output of a particular camera channel.
chiaroscuro The most familiar pictorial treatment, in which a three-dimensional illusion is created by lighting and staging. An impression of solidity and depth is achieved by tonal gradation, carefully related brightnesses of planes, tonal separation, and shadow formations.
chroma See SATURATION.
chroma key (114) An automatic electronic switching (keying) circuit, enabling the subject shot by one camera to be inserted into the corresponding part of another picture. A colour backing (blue, yellow, green) behind the subject activates the switch. Also called *colour separation overlay* (CSO).
CID lamp (20) Compact iodide daylight lamp. A type of gas discharge lamp, available in single-ended bare-bulb and sealed-beam PAR versions. See CSI LAMP.
colour balance (132) The adjustment of the relative strengths of the primaries of a colour system (usually red, green, blue; or cyan, magenta, yellow) to provide optimum colour reproduction fidelity. When a system is accurately balanced throughout, each grey-scale step (value) from 'white' down through greys to 'black', will appear *neutral* (i.e. without colouration). When this *grey-scale tracking* is correct, no colour-bias or colour cast will exist.
colour correction filter (light conversion filter, colour compensating filter) (18) Colour medium placed over a light source to modify its COLOUR TEMPERATURE (colour quality), e.g. to match incandescent light to the colour temperature of daylight.
colour medium (gel) A coloured transparent or translucent material (typically gelatin, acetate, polyester, acrylic sheet) placed in front of a lamp to colour the emergent light. Commercial types of polyester materials include *Cinemoid, Roscolux, Geletran*.
colour separation overlay (CSO) (114) See CHROMA KEY.
colour temperature (16, 18, 128) The colour quality of light, theoretically relating its spectral distribution to that of a heated standard 'black body' radiator. Measured in degrees kelvin. 0 K = −273°C. Each form of luminant tends to have its particular colour quality. Photoflood lighting 3400 K; tungsten-halogen 3200 K; typical daylight 5600 K. The effective colour temperature can be altered by attaching a suitable filter to the light source or the camera lens (bluish filter raises it; yellowish-orange filter lowers it). Such filters made of glass or gelatin are classified in MIREDS (a million divided by the kelvin value), and produce a colour shift dependent on the original colour temperature.
compact iodide daylight lamp See CID LAMP.
compact source iodide lamp (20, 108) See CSI LAMP.
contactor (116) A remotely-operated heavy-duty relay used to switch on/off lamps connected to it.
contrast range (subject brightness range) (12, 128) The brightness ratio between the lightest and darkest tones in a scene. Or the extreme

ratio that a system can accommodate while still reproducing all intermediate tones reasonably accurately.

contrast ratio (12, 128) A measurement of the relative brightness of any two tones, given as a luminance ratio. See BRIGHTNESS.

cookie (50, 100) An irregularly-shaped stencil sheet placed in front of a spotlight to dapple the light beam, to produce either intensity variations, or a distinct shadow pattern.

cross fade (116) Progressively fading out one lamp, while fading up another, usually so that the average light intensity is reasonably constant.

crush-out (12, 24) See BLOOM.

CSI lamp (20, 108) A gas discharge light source comprising two metal electrodes set in a suitable gas vapour. The illumination from a mercury-vapour discharge is colour-improved by the inclusion of a sodium, thallium, gallium filling. It has the disadvantage, however, that it requires corresponding 'warm-up' and 'recovery' periods after switch-on or re-lighting (e.g. 30 secs. to 10 mins. respectively). Single-ended bulb and sealed-beam versions have a colour temperature equal to about 4000 K; a 1 kW version with a 1000-hour life has an output five times that of an overrun tungsten lamp of similar rating. These *metal-halide discharge lamps* are used in projectors, follow spots, effects projectors, flood lighting, to provide high-density, high-efficiency, point-source illumination.

cyc-light (102, 123) A light fitting with a specially-shaped reflector producing a broad elongated light-beam, enabling a background (cyclorama, backing) to be lit evenly overall from a relatively close distance (i.e. acute illumination angle).

cyclorama (cyc) (102) A stretch of taut vertical cloth used as a general-purpose scenic background.

depth of field (24, 76) When a camera is focused sharply on a subject, a zone nearer and further from it still shows detail reasonably clearly. This distance range is known as the *depth of field* (often wrongly called 'depth of focus'). This depth-zone increases as the lens aperture is reduced, or the lens angle is widened, or the focusing distance is increased.

diffuser (scrim) (48, 110) Translucent material attached to the front of a lamp to disperse and soften the light quality, and to reduce its intensity. Spun-glass sheet, frosted plastics sheeting, wire gauze, are widely used.

dimmer (44, 48, 66) An electrical circuit regulating the current flowing through connected lamps, so adjusting their intensities. The solid-state silicon-controlled rectifier (SCR) or thyristor dimmer is increasingly used, replacing bulky, less efficient resistance dimmers. See SCR DIMMER.

dimmer truck (116) A mobile wheeled box containing a series of dimmers.

dipping The process of dyeing a material very slightly (generally coffee, or blue tints) to reduce its reflectance or modify its colours. This is usually necessary to bring the brightness of 'white' fabrics (from shirts to bed linen) within the tonal range of the TV system. Bright white materials will BLOOM and reproduce as detailless areas.

discharge lamp (22, 108) A lamp in which an electric discharge takes place between metal electrodes within a high-pressure gas (e.g. xenon). Used as a substitute for the carbon arc in medium power film and rear-projection equipment. See COMPACT SOURCE IODIDE (CSI) LAMP, XENON LAMP.
dispersion (16) The extent to which light rays are scattered or diffused. Where they emerge from a light source dispersed in all directions, the resultant illumination is 'soft', casting no shadows.
doubling up (98, 124) This can arise when an area already lit by one lamp is inadvertently illuminated by a second, thereby creating multiple shadows, uneven light-coverage, hotspots, etc. Usually avoided, we may occasionally introduce doubling deliberately, where sufficient light intensity is not available from a single light source.
downstage A position near the camera. The opposite of UPSTAGE.
drop-arm (28, 122) A vertical support tube enabling a lamp to be hung from its lower end, some distance below an overhead clamping-point (e.g. at an adjustable or fixed height relative to a ceiling grid). A type of 'hanger'.
dual-source lamp (122) A light fitting that can be transformed from a SOFT-LIGHT source to a FRESNEL spotlight, by electrical or mechanical switching. In more sophisticated designs, two separate source-types are fitted at either end of the same box-housing.

effects spotlight See ELLIPSOIDAL SPOTLIGHT.
ellipsoidal spotlight (profile spot, projection spotlight, effects spot) (21, 23, 82, 92, 123) A spotlight in which the light collected from an ellipsoidal reflector (mirror) is focused by a lens. The shape of its light beam (hard- or soft-edged) is adjustable by an internal variable iris, silhouette mask, or independent internal framing-shutters. It may be designed to project a perforated metal stencil (mask) or a glass slide's image.
environmental effect (94) Lighting and scenery creating the illusion of a particular type of location, at a certain time of day. The lighting simulates an appropriate form of illumination (e.g. candlelight, gaslight, fluorescents, firelight, etc.).
exposure (10, 24, 132) The selective control of reproduced tonal values within a system's limits. Strictly, *over-exposure* results when reflected light exceeds the camera-tube's handling limits. *Under-exposure* arises when a surface is insufficiently bright to be clearly discerned in the picture. However, a subject is often said to be over- or under-exposed when it appears brighter or darker in the picture than is artistically desirable – even where tones lie well within the system's range. Exposure can be controlled by light-intensity adjustment, lens aperture, neutral-density filters, or the camera-tube's target voltage.
external reflector lamp See REFLECTOR SPOTLIGHT.
eyelights (catchlights) (52) Specular reflections in the eyes from frontal

lighting (key light, or camera light), giving a vitality to the portraiture. Multiple eyelights are best avoided.

fill-light (filler, fill-in) (22, 28, 32, 36, 38, 40, 54) Shadowless SOFT LIGHT introduced to reduce the density of shadows cast by the key light (lowering lighting contrast) and to reveal detail in shadows.
filters, colour compensating/correcting (18) See COLOUR CORRECTION FILTER.
flag (50) A metal plate (GOBO) fixed in an adjustable angle-arm, and held in a light beam; to provide a hard- or soft-edged shadow, or to keep light off a selected subject area.
flood (22) A type of light fitting providing illumination over a wide angle; often of open-reflector construction. *Also*, the adjustment of a spotlight to its maximum coverage or spread (usually around 60°).
floor lamp (28) General nomenclature for a lamp fixed to an adjustable tubular metal stand, usually supported by a castored tripod base. The term is sometimes used to denote *ground lamps* resting on the floor.
following spotlight (108) A high-intensity projection spotlight providing a circular beam-shape (hard- or soft-edged) of adjustable diameter. Used to localise illumination, or to place the subject in a pool of light.
foundation light (32) See BASE LIGHT.
fluorescent lamp (12, 20) A tubular lamp in which a mercury-vapour discharge energises a fluorescent powder material coating the inside of the tube. The colour quality of the illumination varies with its constituents.
framing shutters (22) Flat internal plates (strips) within a projector spotlight, enabling the light beam to be selectively shuttered-off and its shape adjusted.
Fresnel lens (stepped lens) (22, 44, 48, 86, 88, 92, 102, 110, 120, 124) The type of lens usually fitted to studio spotlights. Its surface comprises a series of concentric ribs of stepped (triangular) cross-section, making the resultant lens thinner, lighter, and more efficient than a solid lens. It is therefore less vulnerable to overheating and cracking.
frontal light (34) Illumination from the camera direction (e.g. 10 to 15° off the lens axis).

gaffergrip (101) A clip fitting (sprung or screw-adjusted) incorporating a lamp socket, enabling a light to be attached to a narrow structure (e.g. door, chair, pole).
gaffer tape (28, 100) A wide plasticised-fabric adhesive tape that has found many productional applications. Used to tape the bases of lightweight lamp fittings to walls, support wall cables, cover floor cables (lighting or sound), prevent movement of furniture or floor coverings, mark floor positions of furniture and performers, etc.
gamma (132) A measurement of the tonal coarseness of subtlety in the reproducing system. A 'high gamma' picture has few half-tone gradations (i.e. tends to 'soot-and-whitewash'). A 'low gamma' picture reproduces slight grey-scale differences distinctly (often over a narrower contrast

range). Unless the overall gamma of a system is unity, relative tonal and colour values are falsely reproduced.
garland (121) A camera-light assembly surrounding the camera lens.
gobo (50, 100, 115, 121) A jet-black screen (wooden or metal sheet) used to hide lamps or cameras from certain viewpoints. Also in the form of FLAGS attached to adjustable rod-clamps used to cast shadows, and so prevent light from illuminating specific areas.
graphics (120) Artwork of all forms: diagrams, photo-stills, maps, decorative illustrations, titling, credits, etc..
grid (28) A horizontal framework grid near the studio roof, providing access to hoists, lighting suspension, etc. A tubular *lighting grid* is used to support lamps, scenery, etc.
ground cove See MERGING COVE.
ground row (102) A series of lamps in the form of *troughs, battens, cyc lights,* laid on the ground (often behind scenery) to illuminate a cyc or other background. *Also,* scenic term for a low vertical scenic plane behind which such lamps can be hidden from the camera.

hard light (16, 22, 30, 48) Highly directional light casting well-defined shadows and revealing surface texture. Produced by very localised light sources. See SOFT LIGHT.
headlamp (29, 40) See CAMERA LIGHT.
high key (32, 46) A picture in which mid to lighter tones predominate, with little or no shadow. See LOW KEY.
highlights The lightest picture tones (usually specular reflections) taken to 'peak white' (i.e. maximum) in the video signal.
HMI lamp (20, 108) Metal-halide lamp (mercury/argon additives) in tubular, double-ended form. Lightweight 200W, and 2 to 5kW versions. 5600K. See CSI LAMP.
hoist (28) A remotely-winched or motor-driven mechanism, controlling a ceiling-located wire cable. The hoisting-hook is attached to suspended lighting equipment (e.g. a BARREL) or used to support scenery (staging), slung monitors, boom cables, etc.
hot (72, 74) Any over-bright area. It may BLOOM, pale-off, or appear undesirably light. It can result from *over-exposure,* over-lighting (i.e. excessive illumination), or from an unsuitably light or shiny surface.
hot-spot (38, 74, 90, 104) An over-bright localised patch of light – often a specular reflection. *Also,* irregular illumination of a plane intended to be evenly lit; e.g. a rear-projection screen, on which a bright central image falls off in intensity towards its edges.
house lights Powerful ceiling lights used to illuminate the studio overall as a general 'working light' (e.g. for rigging, setting staging, etc.). House lights are extinguished when the specific, controlled lighting equipment is in use.
hue (10, 76, 118) The predominant sensation of colour, i.e. red, green, blue, etc.
hydrargym/medium arc-length/iodide lamp See HMI LAMP.

intensity See SATURATION.
internal-reflector lamp (20, 50) See SEALED-BEAM LAMP.

jelly (44, 118) A plastics medium placed in front of a lamp to reduce light intensity and/or soften the emergent light. Also colour sheeting (e.g. Cinemoid), used to produce coloured light (18, 118).

key light (26, 32, 34, 36, 38, 54, 74) The main light illuminating the subject (usually a FRESNEL spotlight). The key light reveals the subject's form, contours, texture, and may suggest the prevailing light direction (e.g. that a subject is lit from a nearby window).
kick-back (74) When a surface is angled so that it reflects incident light directly back to the camera lens, an over-bright (over-exposed) area may result, preventing detail and texture from being discerned (HOT SPOT, BLOOMING).

lag (trailing) (24) A persistence or smearing after-image following a moving object. When operating with insufficient light (or low beam current) camera tubes are liable to produce this ectoplasmic trailing on movement, especially where light-toned subjects are seen against a dark background.
lamp A general term for an incandescent light source (bulb, 'bubble'). Also used as a loose term for any complete LIGHT FITTING.
lens axis (120) An imaginary line from the middle of a lens to the centre of the shot.
lens flare (42) A spurious blob or streak in the picture, resulting from a bright light or reflection shining into the camera lens, and being internally reflected. Avoided by barndooring the backlight off the camera, raising the backlight, and ensuring that an efficient lens-hood (sun-shade) is fitted to the lens. See BARNDOORS, BACKLIGHT.
lensless reflectors (21) See REFLECTOR SPOTLIGHT.
lens spot (122) A spotlight in which the light reflected by a mirror (often parabolic) is focused by a single-lens system. Adjustment of the lamp/lens distance alters the spotlight's beam angle (spread).
light fitting (22) An enclosure housing a lamp, reflector (mirror), and any associated lens systems. Terms used for such a fitting include: *luminaire, lantern, lamp,* or simply 'light'.
light level (24, 128) The incident light intensity falling upon the scene (in lux, foot-candles) from a light source.
lightness The perceived brightness or surface colour.
light units *Incident light intensity* is measured in *lux (lx)* – formerly *foot-candles* (fc).
Reflected light intensity or *surface brightness* is measured in *nits* (*candela* per sq. m.) – formerly *foot-lamberts*.
Colour temperature (18) i.e. colour quality, is measured in *kelvins* (K).
Light source strength is measured in *lumens* (lm) (candela, cd) – formerly *candles*.
Hue (colour) is measured in *nanometres, ångstroms* (Å) – formerly *millimicrons*.

lighting balance See BALANCE (OF LIGHTING).
lighting plot (122, 126) A scale plan diagram indicating the positions and types of lamps used to light a situation. Details of cabling, patching, loads, accessories, colour media, heights, etc., may be included.
limbo A staging technique in which the subject is seen isolated against a totally white background. The term is often used erroneously, to indicate a neutral or unrecognisable background behind subjects.
low key (116) A picture in which mid to lowest tones predominate with few highlight details. See HIGH KEY.
luminance The true measured brightness of a surface. Doubling the illumination produces double the surface luminance. Snow has a high luminance, black velvet an extremely low luminance. See BRIGHTNESS and LIGHTNESS.
luminosity Alternative term for BRIGHTNESS.

matching (132) Maintaining visual continuity between consecutive pictures, to ensure that they have compatible contrast, brightness, contents, etc.
merging cove (ground cove) (102) A concave ramp arranged at the foot of a cyc or similar background. This enables the floor to appear on camera as merging imperceptibly with the vertical plane, without a visible join or demarcation.
metal halide lamp See HMI LAMP.
mired units (micro reciprocal degrees) Derived by dividing one million by the kelvin value of a colour temperature. (Hence, 5000 K equals 200 mireds). Mireds are used to relate or compare COLOUR TEMPERATURES, especially when selecting compensatory colour filters.
monitor (126) A picture display unit (without an audio output) on which a video output can be continually monitored (appraised) for productional or engineering check purposes. It may be used in the studio to show the production in its entirety (e.g. for cueing, continuity, or audience purposes) or to show non-studio picture-sources (film, slides, video-tape). In the *production control area* (the nerve centre of studio operations) each picture source has its associated preview (channel) monitor. A line monitor (transmission, studio output) shows the studio's switched output (i.e. after the video switching console).
monochrome (10, 18, 76, 114, 118, 132) Strictly speaking, 'single colour'. Customarily refers to 'black-and-white' picture reproduction.

neutral density filter (44, 132) A grey-tinted glass filter of specified density, used to reduce the effective image brightness (for exposure compensation) without affecting colour reproduction or relative tonal values.
noise, video (snow) (24) See VIDEO NOISE.
nook light (101) A small open-fronted trough fitting containing a short strip-light with a curved rear reflector. This lightweight source can be

attached by GAFFERGRIP or GAFFER TAPE to most surfaces. The light quality is fairly hard, but can be used to fill local shadow areas.
notan A pictorial style in which one is concerned with surface detail and colour, and outline, rather than an illusion of solidity and depth.

off-stage A position outside the acting area – often beyond the confines of a setting.
on-stage A position nearer to the centre of the acting area.
open-bulb reflector See REFLECTOR SPOTLIGHT.
opening up (24) The converse of STOPPING DOWN the lens. See EXPOSURE.
overload (134) To exceed the handling capacity of a system. In lighting, an excess power demand relative to a cable or supply's current/wattage rating, causing heating in wiring, connectors, etc., and voltage losses.

parallel See ROSTRUM.
PAR light (20) A tungsten-halogen (quartz) lamp in which its *parabolic aluminised reflector* forms part of the bulb. The internally silvered reflector together with the ribbed moulded-lens front glass provide a fixed beam-spread with narrow (spot), medium, or wide coverage. A blue dichroic filter (for daylight correction) may be incorporated, or clipped on. PAR bulbs are used individually, or grouped in floodlight banks.
patching (122) Each lamp plugs into a wall-outlet socket attached to a supply-line. This line normally terminates in a *patch-cord,* which can be plugged into any of a series of independent numbered supply circuits in a *patchboard (patch panel).* Each power circuit has its associated dimmer and switch control circuits.
Pepper's ghost (114, 120) A method of directing light along the lens axis, for shadowless illumination.
phase (132, 136) Industrial electrical power supplies utilise a *three-phase* alternating current system. As studio lighting equipment requires only a *single-phase* supply, the three incoming supply phases are often separated and distributed, in the form of single-phase supplies, within a studio. Because a high voltage, exceeding the line supply, exists between phases, it is essential as a safety measure, to ensure that neighbouring equipment is fed from the same supply phase (i.e. blue, white or red).
picture tube The cathode-ray tube (CRT) in a TV monitor or receiver, upon the face of which the TV picture is displayed.
pipe grid (28) See GRID.
Plumbicon (129) A lead-oxide camera tube widely used in colour TV cameras. (*Philips* trademark).
pole, lighting (lighting hook) (124) An extensible pole with an end fitting (hook or cup) enabling a suspended luminaire to be controlled from floor level while standing 3 to 6 m/10 to 20 ft below. Adjustments include focus, tilt, pan, barndoors, source switching (hard/soft), power selection. (Remote motor-control methods are also found.)

power rails (28) A ceiling trackway into which suspended lamp fittings can be slotted or hooked, to provide power pick-up without individual lamp cabling.
practical lamps (94, 136) Decorative or environmental light fittings appearing illuminated within a scene, e.g. table lamps.
projection spotlight (profile spotlight) (22, 82, 92, 108) A special projector producing a hard-edged disc of light. Internal diaphragms, cut-out masks, shutters or silhouettes can be used to vary the outline and shape of the light beam. (Metal stencil masks can often be projected.)
purity See SATURATION.

quartz light (20) See TUNGSTEN-HALOGEN LAMP.

rear projection (back projection) (114) A staging arrangement in which action is shot in front of a large translucent sheet, onto which pictures, light-patterns, or shadows are projected from the reverse side.
reflector board (112) A plain white (or metallic) board used to reflect light on location (reflected sunlight as a key or fill-light), or in the studio to provide large-area diffuse illumination.
reflector spotlight (20, 21) A spotlight incorporating a simple parabolic mirror reflector (but no lens), the lamp's beam-angle being adjusted by alteration of the lamp/mirror distance. Widely termed 'external reflector', 'lensless', or 'open-bulb' spotlight, these compact lightweight units are increasingly used on location.
reflex projection (114) A scenic process in which the background image is front-projected along the lens axis onto a special glass-beaded screen.
rig (28, 122) (verb) To install, and set up lamps in their required positions. (noun) The finished assembly of lamps, positioned and patched for a production.
rostrum (parallel) (106) A scenic platform. Used to provide raised flooring, podiums, multi-level staging, etc. of various heights.

safety bond (136) A clip-on wire strop or metal clip attached to equipment, to ensure that accessories cannot become detached, or fall to the ground from suspended lighting equipment.
saturation (chroma, intensity, purity) The purity of colour. As a hue becomes paled-off or diluted by the addition of white light, it becomes *de*-saturated.
scoop (21, 22, 50, 122) An open-fronted soft-light fitting. A spun aluminium reflector (15 to 18 in diameter) housing a frosted lamp of 500 to 2000 W rating. SPILL-RINGS may be fitted.
scratch-off slide (22) A technique in which a black-coated glass slide used for projection is scratched-off to produce clear stencilled sections.
SCR dimmer (silicon-controlled rectifier, thyristor) (44) A solid-state electronics device used to control lamp brightness, by cutting off part of the cycle of the alternating current supply. 'Clean-up chokes' are required to prevent sound-system interference and audible noise from lamps.

scrim (110) A diffuser (48); often of cloth, supported some distance in front of a light source to soften illumination. Term also used for a thin netting (gauze) used to provide translucent cycloramas, or create scenic diffusion.
sealed-beam (internal reflector) lamp (20, 50) A tungsten lamp (running normally, or over-run for higher light output) formed with a specially-shaped bulb or envelope. Its internally-silvered surface serves as a reflector, so that no additional housing is required, save to protect the lamp. See PAR LIGHT.
side light (34) Lighting across the subject, emphasising texture and surface contours. Generally exaggerates modelling; produces poor bisected portraiture on full-face shots.
silhouette (82, 92, 116) A pictorial style which concentrates on subject outline for its effect. Surface detail, tone, texture, colour, are suppressed.
simultaneous contrast (spatial induction) (68) An illusion whereby the apparent tone or hue of a subject is influenced by those of its immediate background or surroundings. Tones look lighter against a dark tone; darker against a light-toned surface.
sit (bat down, set down) (132) Electrical adjustment of the video waveform (its d.c. component) to move all picture tones down (sit down) or up (sit up) the tonal scale. The effect is most marked in darker tonal values.
snoot (50, 108) A conical or cylindrical light-shield fixed to a spotlight for considerably restricting its coverage.
soft light (16, 22, 32, 48) Dispersed scattered light rays producing diffuse, shadowless illumination. Produced by large-area sources, through diffusion media (frosted gels), internal reflection, or multi-lamp sources. See HARD LIGHT.
specular reflections (14, 72, 74, 120) High-intensity reflections of luminants in smooth, shiny, surfaces (glass, metal, plastics, polished wood, gloss paint, etc.). These reflections appear as white blank undetailed areas.
spill light (fresh light, leak light) (72, 96, 104) Light escaping from its intended path, causing spurious patches, streaks, etc., as it falls on nearby surfaces.
spill rings (50) An assembly of shallow concentric cylinders fitted to the front of a light source to reduce light spread. A lattice of vertical and horizontal strips may be similarly used.
spot bar (29, 40) A short bar holding one or two battery-powered lamps, attached to the top of a camera to provide portable illumination. A camera light.
staging (10) Scenery. Scenic treatment in a production, arranged for decorative or environmental effect.
stepped lens See FRESNEL LENS.
stopping down (24) Reducing the camera lens aperture (irising down) to prevent over-exposure of lighter scenic tones, to increase the *depth of field*, or to reduce reproduced tonal values. The converse of OPENING UP the lens. See EXPOSURE.

streaking (24) Well-defined light or dark horizontal stripes superimposed across a picture, resulting from an area of extremely high tonal contrast in the scene – e.g. white venetian blinds, specular reflections, etc.
surface brightness (128) The light reflected from a surface (see LIGHT UNITS) is affected by the relative light and camera angles, incident light intensity, surface finish, the relative colours of the surface and lighting.

target (132) The electric storage surface in a camera tube, on which the scene is focused and a corresponding electron image formed. The voltage applied to the target is adjustable, and affects the picture's contrast (and shading) for a given light level on the scene. Target voltage is often varied as an 'exposure' control with vidicon tubes. A high value increases the camera tube's susceptibility to 'image retention'.
three-point lighting (36, 58, 61, 66, 72) A basic lighting arrangement in which a hard *key light* provides the main illumination, a soft light *filler* (*fill-light*) reveals shadow detail and reduces lighting contrast, and a hard *backlight* reveals outline detail and contour modelling.
throw (108) The distance between a lamp and its destination. Hence a 'long throw' for a distant lamp; a 'short throw' for a close lamp.
transmission monitor (line monitor) (126) The picture monitor showing the 'on-air' programme, i.e. the video source selected on the *production switcher* (*vision mixer*) from those available, for recording or transmission.
trimming (46) General term for final adjustment of a light source's exact coverage and intensity, to produce the required artistic effect.
tungsten-halogen lamp (quartz-iodine/quartz light) (20, 60) A type of lamp design achieving a longer working life, almost constant output and colour temperature and a higher light output. These result from the introduction of a halogen vapour (e.g. bromine, iodine) facilitating chemical re-cycling action; so preventing the substantial filament loss through evaporation that arises in normal *tungsten lamps*. Quartz bulbs should not be touched by hand (body acids attack the glass during operation). Often termed 'quartz lamp'.
tungsten lamp (20) The familiar 'light bulb' in which a coiled tungsten-wire filament heats as an electric current passes through it. Its filament gradually evaporates (often as a black deposit on the glass envelope), its light output and colour temperature falling with use.
twin-filament lamp A lamp containing two separate filaments within the same bulb (envelope). It thus provides full or half-power output by single or double filament switching. This avoids the lowering of colour temperature that results when high-output lamps are *dimmed* to produce low-output illumination.

under-lighting (40, 52, 68, 86) Insufficient illumination relative to the required exposure of lighting balance – viz. under-lit conditions. *Also*, lighting from below the lens axis, usually for horrific or highly dramatic effects; or, more subtly, to nullify modelling or shadows caused by steep lighting.

upstage Strictly, a position towards the back wall of a setting. *Also*, loosely, to indicate a position further away from the camera. Opposite of DOWNSTAGE.

video control (10, 132) Adjustment of various video equipment parameters to control picture quality; including exposure, black level (lift, sit, set-up), video gain, gamma, colour balance. The video operator (vision control) continually monitors the output of all video sources, adjusting them for optimum quality and matching.

video noise (snow) (24) Spurious background scintillations visible throughout a picture; most pronounced in darker areas. Due to random fluctuations in the video signal. (The video equivalent of photographic grain, tape hiss, disc surface noise.)

video tape (114) A magnetic recording system providing instant replay of the television picture and its accompanying sound. The composite tape recording can be edited by *dubbing* (re-recording) or, rarely nowadays, by *splicing* (cutting).

vidicon (24, 129) A TV camera tube widely used in small video cameras. Cheap and reliable, simply adjusted. But of relatively low sensitivity, suffering from *lag*, and spurious background shading.

walkway (28) See CATWALK.

xenon lamp (22, 108) A compact source discharge lamp (often water-cooled) containing xenon gas. Substituting for carbon arcs, its light is of good colour quality, and is primarily used in film and rear-projection equipment. See CSI LAMP.